D1007751

Getting Started with Arduino

Massimo Banzi and Michael Shiloh

SEBASTOPOL, CA

Getting Started with Arduino

by Massimo Banzi and Michael Shiloh

Printed in the United States of America.

Published by Maker Media, Inc., 1005 Gravenstein Highway North, Sebastopol, CA 95472.

Maker Media books may be purchased for educational, business, or sales promotional use. Online editions are also available for most titles (*http://www.safaribooksonline.com*). For more information, contact our corporate/institutional sales department: 800-998-9938 or *corporate@oreilly.com*.

Editor: Brian Jepson
Production Editor: Nicole Shelby
Copyeditor: Kim Cofer
Proofreader: Sharon Wilkey
Indexer: WordCo Indexing Services
Interior Designer: David Futato
Cover Designer: Brian Jepson
Illustrator: Judy Aime' Castro

December 2014:   Third Edition

Revision History for the Third Edition

2014-12-09:   First Release

See *http://oreilly.com/catalog/errata.csp?isbn=9781449363338* for release details.

978-1-449-36333-8

[M]

Contents

Preface

The third edition of *Getting Started with Arduino* adds two new chapters: Chapter 8 is an ambitious project, which illustrates a more complex circuit and program. This chapter also talks about project design, testing, and construction, and makes use of schematic diagrams, which were (and still are) described in Appendix D.

The second chapter, Chapter 6, introduces the Arduino Leonardo. The Leonardo is a different sort of Arduino, because the USB controller is implemented in software, and not in a separate chip as had been the case prior to the Leonardo. This allows the USB behaviour of the board to be modified.

Apart from these new chapters, other updates have taken place:

The Third Edition is written for version 1.0.5 of the IDE. In anticipation of the imminent release of version 1.5, differences between 1.0.5 and 1.5 have been noted.

Numerous suggestions from students and readers have been incorporated.

In keeping with the spirit of the original text, British spelling is used throughout.

—Michael

Preface to the Second Edition

A few years ago I was given a very interesting challenge: teach designers the bare minimum in electronics so that they could build interactive prototypes of the objects they were designing.

I started following a subconscious instinct to teach electronics the same way I was taught in school. Later on I realised that it simply wasn't working as well as I would like, and started to remember sitting in a class, bored like hell, listening to all that theory being thrown at me without any practical application for it.

In reality, when I was in school I already knew electronics in a very empirical way: very little theory, but a lot of hands-on experience.

I started thinking about the process by which I really learned electronics:

- I took apart any electronic device I could put my hands on.
- I slowly learned what all those components were.
- I began to tinker with them, changing some of the connections inside of them and seeing what happened to the device: usually something between an explosion and a puff of smoke.
- I started building some kits sold by electronics magazines.
- I combined devices I had hacked, and repurposed kits and other circuits that I found in magazines to make them do new things.

As a little kid, I was always fascinated by discovering how things work; therefore, I used to take them apart. This passion grew as I targeted any unused object in the house and then took it apart into small bits. Eventually, people brought all sorts of devices for me to dissect. My biggest projects at the time were a dishwasher and an early computer that came from an insurance office, which had a huge printer, electronics cards, magnetic card readers, and many other parts that proved very interesting and challenging to completely take apart.

After quite a lot of this dissecting, I knew what electronic components were and roughly what they did. On top of that, my house was full of old electronics magazines that my father must have bought at the beginning of the 1970s. I spent hours reading the articles and looking at the circuit diagrams without understanding very much.

This process of reading the articles over and over, with the benefit of knowledge acquired while taking apart circuits, created a slow, virtuous circle.

A great breakthrough came one Christmas, when my dad gave me a kit that allowed teenagers to learn about electronics. Every component was housed in a plastic cube that would magnetically snap together with other cubes, establishing a connection; the electronic symbol was written on top. Little did I know that the toy was also a landmark of German design, because Dieter Rams designed it back in the 1960s.

With this new tool, I could quickly put together circuits and try them out to see what happened. The prototyping cycle was getting shorter and shorter.

After that, I built radios, amplifiers, circuits that would produce horrible noises and nice sounds, rain sensors, and tiny robots.

I've spent a long time looking for an English word that would sum up that way of working without a specific plan, starting with one idea and ending up with a completely unexpected result. Finally, *tinkering* came along. I recognised how this word has been used in many other fields to describe a way of operating and to portray people who set out on a path of exploration. For example, the generation of French directors who gave birth to the Nouvelle Vague were called the *tinkerers*. The best definition of tinkering that I've ever found comes from an exhibition held at the Exploratorium in San Francisco:

> Tinkering is what happens when you try something you don't quite know how to do, guided by whim, imagination, and curiosity. When you tinker, there are no instructions—but there are also no failures, no right or wrong ways of doing things. It's about figuring out how things work and reworking them.
>
> Contraptions, machines, wildly mismatched objects working in harmony—this is the stuff of tinkering.
>
> Tinkering is, at its most basic, a process that marries play and inquiry.

From my early experiments I knew how much experience you would need in order to be able to create a circuit that would do what you wanted, starting from the basic components.

Another breakthrough came in the summer of 1982, when I went to London with my parents and spent many hours visiting the Science Museum. They had just opened a new wing dedicated to computers, and by following a series of guided experiments, I learned the basics of binary math and programming.

There I realised that in many applications, engineers were no longer building circuits from basic components, but were instead implementing a lot of the intelligence in their products using microprocessors. Software was replacing many hours of electronic design, and would allow a shorter tinkering cycle.

When I came back, I started to save money, because I wanted to buy a computer and learn how to program.

My first and most important project after that was using my brand-new ZX81 computer to control a welding machine. I know it doesn't sound like a very exciting project, but there was a need for it and it was a great challenge for me, because I had just learned how to program. At this point, it became clear that writing lines of code would take less time than modifying complex circuits.

Twenty-odd years later, I'd like to think that this experience allows me to teach people who don't even remember taking any math class and to infuse them with the same enthusiasm and ability to tinker that I had in my youth and have kept ever since.

—Massimo

Acknowledgments for Massimo Banzi

This book is dedicated to Ombretta.

Acknowledgments for Michael Shiloh

This book is dedicated to my brother and my parents.

First of all I'd like to thank Massimo for inviting me to work on the Third Edition of this book, and for inviting me to join Arduino in general. It's been a real privilege and joy to participate in this project.

Brian Jepson for guidance, oversight, encouragement, and support. Frank Teng for keeping me on track. Kim Cofer and Nicole Shelby for doing a wonderful job of copyediting and production editing, respectively.

My daughter Yasmine for thinking so highly of me, for her never-ending support and encouragement of my pursuing my interests, and for still thinking that I'm kinda cool in spite of being her dad. I could not have done this without her support.

Last but not least, my partner Judy Aime' Castro for the endless hours she spent turning my scribbles into fine illustrations, for discussing various aspects of the book, and for her endless patience with me. I could not have done this without her support either.

Conventions Used in This Book

The following typographical conventions are used in this book:

Italic
: Indicates new terms, URLs, email addresses, filenames, and file extensions.

`Constant width`
: Used for program listings, as well as within paragraphs to refer to program elements such as variable or function names, databases, data types, environment variables, statements, and keywords.

`Constant width bold`
: Shows commands or other text that should be typed literally by the user.

`Constant width italic`
: Shows text that should be replaced with user-supplied values or by values determined by context.

 This icon signifies a tip, suggestion, or general note.

This icon indicates a warning or caution.

Using Code Examples

This book is here to help you get your job done. In general, you may use the code in this book in your programs and documentation. You do not need to contact us for permission unless you're reproducing a significant portion of the code. For example, writing a program that uses several chunks of code from this book does not require permission. Selling or distributing a CD-ROM of examples from Make: books does require permission. Answering a question by citing this book and quoting example code does not require permission. Incorporating a significant amount of example code from this book into your product's documentation does require permission.

We appreciate, but do not require, attribution. An attribution usually includes the title, author, publisher, and ISBN. For example: "*Getting Started With Arduino, Third Edition*, by Massimo Banzi and Michael Shiloh (Maker Media). Copyright 2015 Massimo Banzi and Michael Shiloh, 978-1-4493-6333-8."

If you feel your use of code examples falls outside fair use or the permission given here, feel free to contact us at *bookpermissions@makermedia.com*.

Safari® Books Online

Safari Books Online is an on-demand digital library that delivers expert content in both book and video form from the world's leading authors in technology and business.

Technology professionals, software developers, web designers, and business and creative professionals use Safari Books Online as their primary resource for research, problem solving, learning, and certification training.

Safari Books Online offers a range of plans and pricing for enterprise, government, education, and individuals.

Members have access to thousands of books, training videos, and prepublication manuscripts in one fully searchable database from publishers like Maker Media, O'Reilly Media, Prentice Hall Professional, Addison-Wesley Professional, Microsoft Press, Sams, Que, Peachpit Press, Focal Press, Cisco Press, John Wiley & Sons, Syngress, Morgan Kaufmann, IBM Redbooks, Packt, Adobe Press, FT Press, Apress, Manning, New Riders, McGraw-Hill, Jones & Bartlett, Course Technology, and hundreds more. For more information about Safari Books Online, please visit us online.

How to Contact Us

Please address comments and questions concerning this book to the publisher:

Make:

1005 Gravenstein Highway North

Sebastopol, CA 95472

800-998-9938 (in the United States or Canada)

707-829-0515 (international or local)

707-829-0104 (fax)

Make: unites, inspires, informs, and entertains a growing community of resourceful people who undertake amazing projects in their backyards, basements, and garages. Make: celebrates your right to tweak, hack, and bend any technology to your will. The Make: audience continues to be a growing culture and community that believes in bettering ourselves, our environment, our educational system—our entire world. This is much more than an audience, it's a worldwide movement that Make is leading—we call it the Maker Movement.

For more information about Make:, visit us online:

Make: magazine: *http://makezine.com/magazine*
Maker Faire: *http://makerfaire.com*
Makezine.com: *http://makezine.com*
Maker Shed: *http://makershed.com*

We have a kit with the components needed to try most of the examples (through the end of Chapter 7) available from the Maker Shed (*http://bit.ly/get-started-arduino-v3*).

We also have a web page for this book, where we list errata, examples, corrections to the code, and any additional information. You can access this page at *http://bit.ly/start_arduino_3e*.

For more information about Arduino, including discussion forums and further documentation, see *http://www.arduino.cc*.

To comment or ask technical questions about this book, send email to: *bookquestions@oreilly.com*.

1/Introduction

Arduino is an open source *physical computing* platform for creating interactive objects that stand alone or collaborate with software on your computer. Arduino was designed for artists, designers, and others who want to incorporate physical computing into their designs without having to first become electrical engineers.

The Arduino hardware and software is open source. The open source philosophy fosters a community that shares its knowledge generously. This is great for beginners as help is often available geographically nearby and always online, at many different skill levels, and on a bewildering array of topics. Example projects are presented not just as pictures of the finished project, but include instructions for making your own or as a starting point for incorporation into your derivative or related projects.

The Arduino software, known as the Integrated Development Environment (IDE), is free. You can download it from *www.arduino.cc*. The Arduino IDE is based on the Processing language (*http://www.processing.org*), which was developed to help artists create computer art without having to first become software engineers. The Arduino IDE can run on Windows, Macintosh, and Linux.

The Arduino board is inexpensive (about $30) and quite tolerant of common novice mistakes. If you do somehow manage to damage the main component on the Arduino Uno, it can be replaced for as little as $4.

The Arduino project was developed in an educational environment and is a very popular educational tool. The same open source philosophy that created the community which generously shares information, answers, and projects also shares teaching methods, curricula, and other information. Arduino has a special mailing list (*http://bit.ly/1vKhOwb*) to facilitate discussion among anyone interested in teaching with or about Arduino.

Because the Arduino hardware and software are open source, you can download the Arduino hardware design and build your own, or use it as a starting point for your own project, based on (or incorporating) Arduino within its design, or simply to understand how Arduino works. You can do the same things with the software.

This book is designed to help beginners with no prior experience get started with Arduino.

Intended Audience

This book was written for the "original" Arduino users: designers and artists. Therefore, it tries to explain things in a way that might drive some engineers crazy. Actually, one of them called the introductory chapters of the first draft "fluff". That's precisely the point. Let's face it: most engineers aren't able to explain what they do to another engineer, let alone a regular human being. Let's now delve deep into the fluff.

This book is not meant to be a textbook for teaching electronics or programming, but you will learn something about electronics and programming while reading this book.

> After Arduino started to become popular, I realised how experimenters, hobbyists, and hackers of all sorts were starting to use it to create beautiful and crazy objects. I realised that you're all artists and designers in your own right, so this book is for you as well.
>
> —Massimo

 Arduino builds upon the thesis work Hernando Barragan did on the Wiring platform while studying under Casey Reas and me (Massimo) at Interaction Design Institute Ivrea (IDII).

What Is Interaction Design?

Arduino was born to teach Interaction Design, a design disci-pline that puts prototyping at the centre of its methodology. There are many definitions of Interaction Design, but the one that we prefer is this:

Interaction Design is the design of any interactive experience.

In today's world, Interaction Design is concerned with the cre-ation of meaningful experiences between us (humans) and objects. It is a good way to explore the creation of beautiful—and maybe even controversial—experiences between us and technology. Interaction Design encourages design through an iterative process based on prototypes of ever-increasing fidelity. This approach—also part of some types of conventional design—can be extended to include prototyping with technology; in particular, prototyping with electronics.

The specific field of Interaction Design involved with Arduino is physical computing (or Physical Interaction Design).

What Is Physical Computing?

Physical computing uses electronics to prototype new objects for designers and artists. It involves the design of interactive objects that can communicate with humans by using sensors and actuators controlled by a behaviour implemented as soft-ware running inside a microcontroller (a small computer on a single chip).

In the past, using electronics meant having to deal with engi-neers all the time, and building circuits one small component at a time; these issues kept creative people from playing around with the medium directly. Most of the tools were meant for engi-neers and required extensive knowledge.

In recent years, microcontrollers have become cheaper and eas-ier to use. At the same time, computers have become faster and more powerful, allowing the creation of better (and easier) development tools.

The progress that we have made with Arduino is to bring these tools one step closer to the novice, allowing people to start building stuff after only two or three days of a workshop. With Arduino, a designer or artist can get to know the basics of electronics and sensors very quickly and can start building prototypes with very little investment.

2/The Arduino Way

The Arduino philosophy is based on making designs rather than talking about them. It is a constant search for faster and more powerful ways to build better prototypes. We have explored many prototyping techniques and developed ways of thinking with our hands.

Classic engineering relies on a strict process for getting from A to B; the Arduino Way delights in the possibility of getting lost on the way and finding C instead.

This is the tinkering process that we are so fond of—playing with the medium in an open-ended way and finding the unexpected. In this search for ways to build better prototypes, we also selected a number of software packages that enable the process of constant manipulation of the software and hardware medium.

The next few sections present some philosophies, events, and pioneers that have inspired the Arduino Way.

Prototyping

Prototyping is at the heart of the Arduino Way: we make things and build objects that interact with other objects, people, and networks. We strive to find a simpler and faster way to prototype in the cheapest possible way.

A lot of beginners approaching electronics for the first time think that they have to learn how to build everything from scratch. This is a waste of energy: what you want is to be able to confirm that something's working very quickly so that you can motivate yourself to take the next step or maybe even motivate somebody else to give you a lot of cash to do it.

This is why we developed *opportunistic prototyping*: why spend time and energy building from scratch, a process that requires time and in-depth technical knowledge, when we can take ready-made devices and hack them in order to exploit the hard work done by large companies and good engineers?

Our hero is James Dyson (*http://www.dyson.co.uk/*), who made 5127 prototypes of his vacuum cleaner before he was satisfied that he'd gotten it right.

Tinkering

We believe that it is essential to play with technology, exploring different possibilities directly on hardware and software—sometimes without a very defined goal.

Reusing existing technology is one of the best ways of tinkering. Getting cheap toys or old discarded equipment and hacking them to make them do something new is one of the best ways to get to great results.

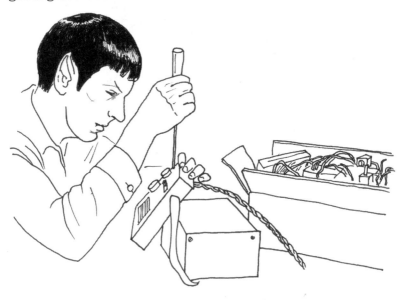

Patching

> I have always been fascinated by modularity and the ability to build complex systems by connecting together simple devices. This process is very well represented by Robert Moog and his analogue synthesizers. Musicians constructed sounds, trying endless combinations by patching together different modules with cables. This approach made the synthesizer look like an old telephone switch, but combined with the numerous knobs, that was the perfect platform for tinkering with sound and innovating music. Moog described it as a process between "witnessing and discovering". I'm sure most musicians at first didn't know what all those hundreds of knobs did, but they tried and tried, refining their own style with no interruptions in the flow.
>
> —Massimo

Reducing the number of interruptions to the flow is very important for creativity—the more seamless the process, the more tinkering happens.

This technique has been translated into the world of software by visual programming environments like Max, Pure Data, or VVVV. These tools can be visualised as *boxes* for the different functionalities that they provide, letting the user build *patches* by connecting these boxes together. These environments let the user experiment with programming without the constant interruption typical of the usual cycle: "type program, compile, damn—there is an error, fix error, compile, run". If you are more visually minded, we recommend that you try them out.

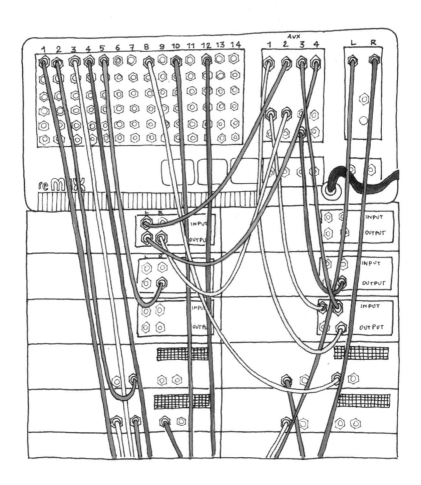

Circuit Bending

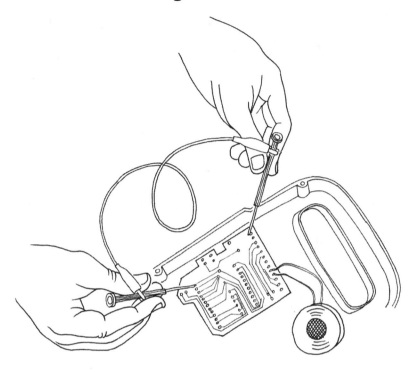

Circuit bending is one of the most interesting forms of tinkering. It's the creative short-circuiting of low-voltage, battery-powered electronic audio devices such as guitar effect pedals, children's toys, and small synthesizers to create new musical instruments and sound generators. The heart of this process is the "art of chance". It began in 1966 when Reed Ghazala, by chance, shorted-out a toy amplifier against a metal object in his desk drawer, resulting in a stream of unusual sounds. Circuit benders excel in their ability to create the wildest devices by tinkering away with technology without necessarily understanding what they are doing on the theoretical side.

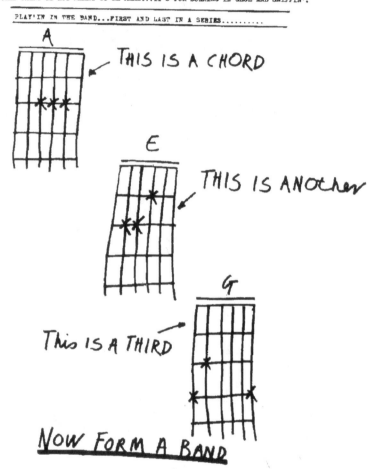

SNIFFIN' GLUE..
+ OTHER ROCK 'N' ROLL HABITS
FOR PUNKS! ① NO.1 OF MANY, WE HOPE!

THIS THING IS NOT MEANT TO BE READ...IT'S FOR SOAKING IN GLUE AND SNIFFIN'.

PLAY'IN IN THE BAND...FIRST AND LAST IN A SERIES..........

A

THIS IS A CHORD

E

THIS IS ANOTHER

G

This IS A THIRD

NOW FORM A BAND

It's a bit like the *Sniffin' Glue* fanzine shown here: during the punk era, knowing three chords on a guitar was enough to start a band. Don't let the experts in one field tell you that you'll never be one of them. Ignore them and surprise them.

Keyboard Hacks

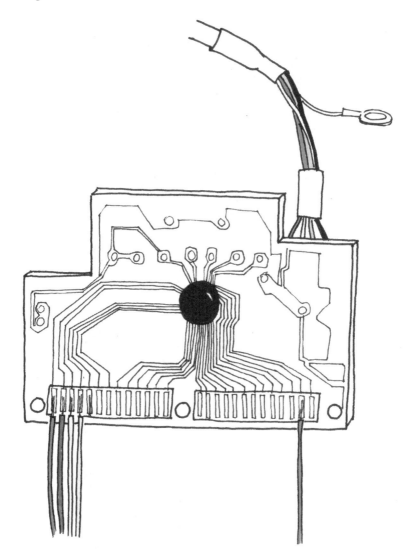

Computer keyboards are still the main way to interact with a computer after more than 60 years. Alex Pentland, academic head of the MIT Media Laboratory, once remarked: "Excuse the expression, but men's urinals are smarter than computers. Computers are isolated from what's around them."[1]

As tinkerers, we can implement new ways to interact with software by replacing the keys with devices that are able to sense the environment. Taking apart a computer keyboard reveals a very simple (and cheap) device. The heart of it is a small board. It's normally a smelly green or brown circuit with two sets of contacts going to two plastic layers that hold the connections between the different keys. If you remove the circuit and use a wire to bridge two contacts, you'll see a letter appear on the computer screen. If you go out and buy a motion-sensing detector and connect this to your keyboard, you'll see a key being pressed every time somebody walks in front of the computer. Map this to your favourite software, and you have made your computer as smart as a urinal. Learning about keyboard hacking is a key building block of prototyping and physical computing.

We Love Junk!

People throw away a lot of technology these days: old printers, computers, weird office machines, technical equipment, and even a lot of military stuff. There has always been a big market for this surplus technology, especially among young and/or poorer hackers and those who are just starting out. This market became evident in Ivrea, where we developed Arduino. The city used to be the headquarters of the Olivetti company. They had been making computers since the 1960s; in the mid 1990s, they threw everything away in junkyards in the area. These are full of computer parts, electronic components, and weird devices of all kinds. We spent countless hours there, buying all sorts of contraptions for very little money and hacking into our prototypes. When you can buy a thousand loudspeakers for very little money, you're bound to come up with some idea in the end. Accumulate junk and go through it before starting to build something from scratch.

---

1 Quoted in Sara Reese Hedberg's "MIT Media Lab's quest for perceptive computers," *Intelligent Systems and Their Applications*, IEEE, Jul/Aug 1998.

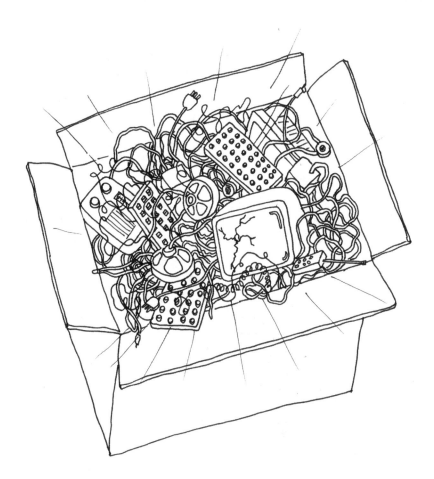

Hacking Toys

Toys are a fantastic source of cheap technology to hack and reuse, as evidenced by the practise of circuit bending mentioned earlier. With the current influx of thousands of very cheap high-tech toys from China, you can build quick ideas with a few noisy cats and a couple of light swords.

> I have been doing this for a few years to get my students to understand that technology is not scary or difficult to approach. One of my favourite resources is the booklet "Low Tech Sensors and Actuators" (*http://lowtech.proposi tions.org.uk*) by Usman Haque and Adam

Somlai-Fischer. I think they have perfectly
described this technique in that handbook, and I
have been using it ever since.

—Massimo

Collaboration

Collaboration between users is one of the key principles in the
Arduino world—through the forum at *www.arduino.cc*, people
from different parts of the world help each other learn about the
platform. The Arduino team encourages people to collaborate at
a local level as well by helping them set up users' groups in
every city they visit. We also set up a wiki called "Playground"
(*http://www.arduino.cc/playground*) where users document
their findings. It's so amazing to see how much knowledge these
people pour out on the Web for everybody to use.

> This culture of sharing and helping each other is
> one of the things that I'm most proud of in
> regard to Arduino.
>
> —Massimo

3/The Arduino Platform

Arduino is composed of two major parts: the Arduino board, which is the piece of hardware you work on when you build your objects; and the Arduino Integrated Development Environment, or IDE, the piece of software you run on your computer. You use the IDE to create a *sketch* (a little computer program) that you upload to the Arduino board. The sketch tells the board what to do.

Not too long ago, working on hardware meant building circuits from scratch, using hundreds of different components with strange names like resistor, capacitor, inductor, transistor, and so on. Every circuit was wired to do one specific application, and making changes required you to cut wires, solder connections, and more.

With the appearance of digital technologies and microprocessors, these functions, which were once implemented with wires, were replaced by software. Software is easier to modify than hardware. With a few keypresses, you can radically change the logic of a device and try two or three versions in the same amount of time that it would take you to solder a couple of resistors.

The Arduino Hardware

The Arduino board is a small microcontroller board, which is a small circuit (the board) that contains a whole computer on a small chip (the microcontroller).

This computer is at least a thousand times less powerful than the MacBook I'm using to write this, but it's a lot cheaper and very useful for building interesting devices.

—Massimo

Look at the Arduino Uno board: you'll see a rectangular black piece of plastic with 28 "legs" (or possibly a tiny square piece of plastic if you have the SMD edition)—that chip is the ATmega328, the heart of your board.

 In fact, there are a variety of Arduino boards, but the most common one by far is the Arduino Uno, which is described here. In Chapter 6 you'll learn about one of the other Arduino boards.

We (the Arduino team) have placed on this board all the components that are required for this microcontroller to work properly and to communicate with your computer. There are many versions of this board; the one we'll use throughout this book is the Arduino Uno, which is the simplest one to use and the best one for learning on. Almost everything we'll talk about applies to all Arduinos, including the most recent ones as well as the earlier ones. Figure 3-1 shows the Arduino Uno.

In Figure 3-1, you see that the Arduino has a row of strips at the top and the bottom with lots of labels. These strips are the connectors, which are used to attach to *sensors* and *actuators*. (An actuator is the opposite of a sensor: a sensor senses something in the physical world and converts it to a signal a computer can understand, while an actuator converts a signal from a computer into an act in the physical world. You'll learn much more about sensors and actuators in this book.)

At first, all those connectors might be a little confusing. Here is an explanation of the input and output pins you'll learn to use in this book. Don't worry if you're still confused after reading this—there are many new concepts in this book that might take you a while to get used to. We'll repeat these explanations a number of different ways, and they'll especially start making sense to you once you start building circuits and experiencing the results.

14 Digital I/O pins (pins 0–13)
These pins can be either *inputs* or *outputs*. Inputs are used to read information from sensors, while outputs are used to control actuators. You will specify the direction (in or out) in the sketch you create in the IDE. Digital inputs can only read one of two values, and digital outputs can only output one of two values (HIGH and LOW).

6 Analogue In pins (pins 0–5)
The analogue input pins are used for reading voltage measurements from analogue sensors. In contrast to digital inputs, which can distinguish between only two different levels (HIGH and LOW), analogue inputs can measure 1,024 different levels of voltage.

6 Analogue Out pins (pins 3, 5, 6, 9, 10, and 11)
These are actually six of the digital pins that can perform a third function: they can provide analogue output. As with the digital I/O pins, you specify what the pin should do in your sketch.

The board can be powered from your computer's USB port, most USB chargers, or an AC adapter (9 volts recommended, 2.1 mm barrel tip, center positive). Whenever power is provided at the power socket, Arduino will use that, and if there is no power at the power socket, Arduino will use power from the USB socket. It's safe to have power at both the power socket and the USB socket.

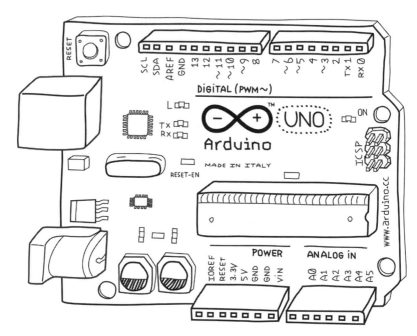

Figure 3-1. *The Arduino Uno*

The Software Integrated Development Environment (IDE)

The IDE is a special program running on your computer that allows you to write sketches for the Arduino board in a simple language modeled after the Processing language (*http://www.processing.org*). The magic happens when you press the button that uploads the sketch to the board: the code that you have written is translated into the C language (which is generally quite hard for a beginner to use), and is passed to the *avr-gcc compiler*, an important piece of open source software that makes the final translation into the language understood by the microcontroller. This last step is quite important, because it's where Arduino makes your life simple by hiding away most of the complexities of programming microcontrollers.

The programming cycle on Arduino is basically as follows:

1. Plug your board into a USB port on your computer.

2. Write a sketch that will bring the board to life.

3. Upload this sketch to the board through the USB connection and wait a couple of seconds for the board to restart.

4. The board executes (performs) the sketch that you wrote.

Installing Arduino on Your Computer

To program the Arduino board, you must first install the IDE by downloading the appropriate file from the *Arduino website* (*http://www.arduino.cc/en/Main/Software*). Choose the right version for your operating system, and then proceed with the appropriate instructions in the following sections.

 See the *"Learning Linux" section on the Arduino web-site* (*http://playground.arduino.cc/Learning/Linux*) for Linux installation instructions.

Installing the IDE: Macintosh

When the file download has finished, double-click to open it, which will open a disk image that contains the Arduino application.

Drag the Arduino application into your *Applications* folder.

Configuring the Drivers: Macintosh

The Arduino Uno uses a driver provided by the Macintosh operating system, so there is nothing to install.

Now that the IDE is installed, connect your Arduino Uno to your Macintosh via a USB cable.

The green LED labeled *PWR* on the board should come on, and the yellow LED labeled *L* should start blinking.

You might see a pop-up window telling you that a new network interface has been detected.

If that happens, Click Network Preferences, and when it opens, click Apply. The Uno will show up as Not Configured, but it's working properly. Quit System Preferences.

Now that you've configured the software, you need to select the proper port to communicate with the Arduino Uno.

Port Identification: Macintosh

Invoke the Arduino IDE, either through the *Applications* folder or by using Spotlight.

From the Tools menu in the Arduino IDE, select Serial Port and then select the port that begins with */dev/cu.usbmodem* or */dev/tty.usbmodem*. Both of these ports refer to your Arduino board, and it makes no difference which one you select.

Figure 3-2 shows the list of ports.

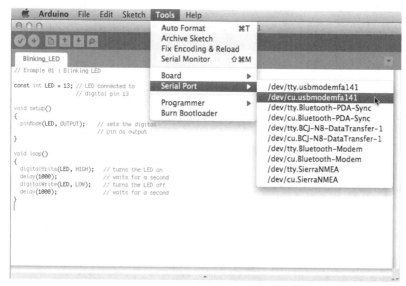

Figure 3-2. *The Arduino IDE's list of serial ports on a Macintosh*

You're almost done! The final thing you should check is that Arduino is configured for the type of board you're using.

From the Tools menu in the Arduino IDE, select Board, and then select Arduino Uno. If you have a different board, you'll need to select that board type (the name of the board is printed next to the Arduino symbol).

Congratulations! Your Arduino software is installed, configured, and ready to use. You're ready to go on to Chapter 4.

 If you have trouble with any of these steps, see Chapter 9, Troubleshooting.

Installing the IDE: Windows

When the file download has finished, double-click to open the installer.

You will be shown a license. Read the license, and if you agree with it, click the I Agree button.

You will be given a list of components to install, and, by default, all of them will be selected. Leave them all selected and click Next.

You will be asked to select an installation folder, and the installer will propose a default for that. Unless you have a good reason not to, accept the default and click Install.

The installer will display its progress as it extracts and installs the files.

After the files are installed, a window will pop up asking for permission to install the drivers. Click Install.

When the installer has completed, click Close to finish.

Configuring the Drivers: Windows

Now that the IDE is installed, connect your Arduino Uno to your computer via a USB cable.

The green LED labeled *PWR* on the board should come on, and the yellow LED labeled *L* should start blinking.

The Found New Hardware Wizard window comes up, and Windows should automatically find the right drivers.

 If you have trouble with any of these steps, see "Problems Installing Drivers on Windows" on page 199 in Chapter 9.

Now that the driver has been configured, you need to select the proper port to communicate with the Arduino Uno.

Port Identification: Windows

Run the Arduino IDE, either using a desktop shortcut or the Start menu.

From the Tools menu in the Arduino IDE, select Serial Port. You will see one or more COM ports with different numbers. Make a note of which numbers are available.

Now unplug your Arduino from your computer, look at the list of ports again, and see which COM port vanishes. It might take a moment or two, and you may have to leave the Tools menu and open it again to refresh the list of ports.

 If you have trouble identifying the COM port used by your Arduino Uno, see "Identifying the Arduino COM Port on Windows" on page 200 in Chapter 9.

Once you've figured out the COM port assignment, you can select that port from the Tools→Serial Port menu in the Arduino IDE.

You're almost done! The final thing you should check is that Arduino is configured for your type of board.

From the Tools menu in the Arduino IDE, select Board and select Arduino Uno. If you have a different board, you'll need to select that board type (the name of the board is printed next to the Arduino symbol).

Congratulations! Your Arduino software is installed, configured, and ready to use. You're ready to go on to Chapter 4.

4/Really Getting Started with Arduino

Now you'll learn how to build and program an interactive device.

Anatomy of an Interactive Device

All of the objects we will build using Arduino follow a very simple pattern that we call the *interactive device*. The interactive device is an electronic circuit that is able to sense the environment by using *sensors* (electronic components that convert real-world measurements into electrical signals). The device processes the information it gets from the sensors with behaviour that's described in the software. The device will then be able to interact with the world by using *actuators*, electronic components that can convert an electric signal into a physical action.

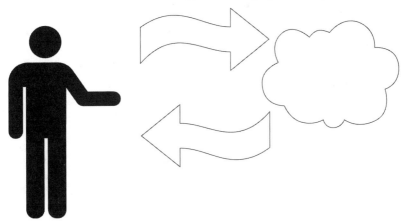

Sensors and Actuators

Sensors and actuators are electronic components that allow a piece of electronics to interact with the world.

As the microcontroller is a very simple computer, it can process only electric signals (a bit like the electric pulses that are sent between neurons in our brains). For it to sense light, temperature, or other physical quantities, it needs something that can convert them into electricity. In our body, for example, the eye converts light into signals that get sent to the brain using nerves. In electronics, we can use a simple device called a *light-dependent resistor* (LDR), also known as a *photoresistor*, that can measure the amount of light that hits it and report it as a signal that can be understood by the microcontroller.

Once the sensors have been read, the device has the information needed to decide how to react. The decision-making process is handled by the microcontroller, and the reaction is performed by actuators. In our bodies, for example, muscles receive electric signals from the brain and convert them into a movement. In the electronic world, these functions could be performed by a light or an electric motor.

In the following sections, you will learn how to read sensors of different types and control different kinds of actuators.

Blinking an LED

The LED blinking sketch is the first program that you should run to test whether your Arduino board is working and is configured correctly. It is also usually the very first programming exercise someone does when learning to program a microcontroller. A *light-emitting diode* (LED) is a small electronic component that's a bit like a lightbulb, but is more efficient and requires a lower voltage to operate.

Your Arduino board comes with an LED preinstalled. It's marked *L* on the board. This preinstalled LED is connected to pin number 13. Remember that number because we'll need to use it later. You can also add your own LED—connect it as shown in Figure 4-1. Note that it's plugged into the connector hole that is labeled 13.

If you intend to keep the LED lit for a long period of time, you should use a resistor as described in "Controlling Light with PWM" on page 56.

K indicates the cathode (negative), or shorter lead; A indicates the anode (positive), or longer lead.

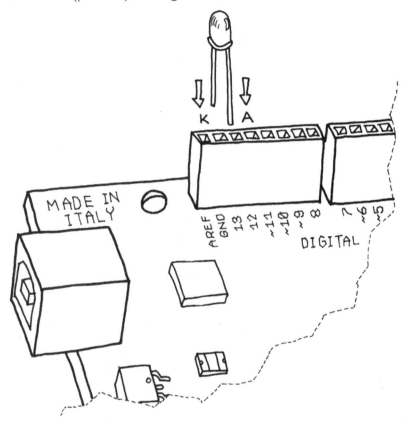

Figure 4-1. *Connecting an LED to Arduino*

Once the LED is connected, you need to tell Arduino what to do. This is done through code: a list of instructions that you give the microcontroller to make it do what you want. (The words *code*, *program*, and *sketch* are all terms that refer to this same list of instructions.)

On your computer, run the Arduino IDE (on the Mac, it should be in the *Applications* folder; on Windows, the shortcut will be either on your desktop or in the Start menu). Select File→New and you'll be asked to choose a sketch folder name: this is where your Arduino sketch will be stored. Name it *Blinking_LED* and click OK. Then, type the following sketch (Example 4-1) into the Arduino sketch editor (the main window of the Arduino IDE). You can also download it from the example code link on the book's catalog page (*http://bit.ly/start_arduino_3e*).

You can also load this sketch simply by clicking File→Examples→01.Basics→Blink, but you'll learn better if you type it in yourself.

It should appear as shown in Figure 4-2.

Example 4-1. Blinking LED

```
// Blinking LED

const int LED = 13; // LED connected to
                    // digital pin 13

void setup()
{
  pinMode(LED, OUTPUT);   // sets the digital
                          // pin as output
}

void loop()
{
  digitalWrite(LED, HIGH);   // turns the LED on
  delay(1000);               // waits for a second
  digitalWrite(LED, LOW);    // turns the LED off
  delay(1000);               // waits for a second
}
```

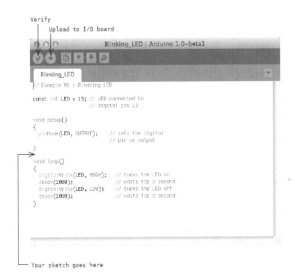

Figure 4-2. *The Arduino IDE with your first sketch loaded*

Now that the code is in your IDE, you need to verify that it is correct. Click the Verify button (Figure 4-2 shows its location); if everything is correct, you'll see the message "Done compiling" appear at the bottom of the Arduino IDE. This message means that the Arduino IDE has translated your sketch into an executable program that can be run by the board, a bit like an *.exe* file in Windows or an *.app* file on a Mac.

If you get an error, most likely you made a mistake typing in the code. Look at each line very carefully and check each and every character, especially symbols like parentheses, braces, semicolons, and commas. Make sure you've copied uppercase and lowercase faithfully, and that you've used the letter O and the number 0 correctly.

Once your code verifies correctly, you can upload it into the board by clicking the Upload button (see Figure 4-2). This will tell the IDE to start the upload process, which first resets the Arduino board, forcing it to stop what it's doing and listen for instructions coming from the USB port. The Arduino IDE will then send the sketch to the Arduino board, which will store the sketch in its permanent memory. Once the IDE has sent the entire sketch, the Arduino board will start running your sketch.

This happens fairly quickly. If you keep your eyes on the bottom of the Arduino IDE, you will see a few messages appear in the black area at the bottom of the window, and just above that area, you might see the message "Compiling," then "Uploading," and finally "Done uploading" to let you know the process has completed correctly.

There are two LEDs, marked RX and TX, on the Arduino board; these flash every time a byte is sent or received by the board. During the upload process, they keep flickering. This also happens very quickly, so unless you're looking at your Arduino board at the right time, you might miss it.

If you don't see the LEDs flicker, or if you get an error message instead of "Done uploading", then there is a communication problem between your computer and Arduino. Make sure you've selected the right serial port (see Chapter 3) in the Tools→Serial Port menu. Also, check the Tools→Board menu to confirm that the correct model of Arduino is selected there.

If you are still having problems, check Chapter 9.

Once the code is in your Arduino board, it will stay there until you put another sketch on it. The sketch will survive if the board is reset or turned off, a bit like the data on your computer's hard drive.

Assuming that the sketch has been uploaded correctly, you will see the LED L turn on for a second and then turn off for a second. If you installed a separate LED as shown back in Figure 4-1, that LED will blink too. What you have just written and run is a *computer program*, or *sketch*, as Arduino programs are called. Arduino, as I've mentioned before, is a small computer, and it can be programmed to do what you want. This is done using a programming language to type a series of instructions in the Arduino IDE, which turns it into an executable for your Arduino board.

We'll next show you how to understand the sketch. First of all, the Arduino executes the code sequentially from top to bottom, so the first line at the top is the first one read; then it moves down, a bit like how the playhead of a video player like Quick-

Time Player or Windows Media Player moves from left to right, showing where you are in the movie.

Pass Me the Parmesan

Notice the presence of curly braces, which are used to group lines of code together. These are particularly useful when you want to give a name to a group of instructions. If you're at dinner and you ask somebody, "Please pass me the Parmesan cheese," this kicks off a series of actions that are summarised by the small phrase that you just said. As we're humans, it all comes naturally, but all the individual tiny actions required to do this must be spelled out to the Arduino, because it's not as powerful as our brain. So to group together a number of instructions, you stick a { before the block of code and a } after.

You can see that there are two blocks of code defined in this way here. Before each one of them are some strange words:

```
void setup()
```

This line gives a name to a block of code. If you were to write a list of instructions that teach Arduino how to pass the Parmesan, you would write `void passTheParmesan()` at the beginning of a block, and this block would become an instruction that you can call from anywhere in the Arduino code. These blocks are called *functions*. Now that you've created a function from this block of code, you can write `passTheParmesan()` anywhere in your sketch, and Arduino will jump to the `passTheParmesan()` function, execute those instructions, and then jump back to where it was and continue where it left off.

This points out something important about any Arduino program. Arduino can do only one thing at a time, one instruction at a time. As Arduino runs your program, line by line, it's *executing*, or running, only that one line. When it jumps to a function, it executes the function, line by line, before returning to where it was. Arduino can't run two sets of instructions at the same time.

Arduino Is Not for Quitters

Arduino always expects that you've created two functions: one called `setup()` and one called `loop()`.

setup() is where you put all the code that you want to execute once at the beginning of your program, and loop() contains the core of your program, which is executed over and over again. This is done because Arduino is not like your regular computer —it cannot run multiple programs at the same time, and programs can't quit. When you power up the board, the code runs; when you want to stop, you just turn it off.

Real Tinkerers Write Comments

Any text beginning with // is ignored by Arduino. These lines are *comments*, which are notes that you leave in the program for yourself, so that you can remember what you did when you wrote it, or for somebody else, so they can understand your code.

It is very common (we know this because we do it all the time) to write a piece of code, upload it onto the board, and say "OK— I'm never going to have to touch this sucker again!" only to realise six months later that you need to update the code or fix a bug. At this point, you open up the program, and if you haven't included any comments in the original program, you'll think, "Wow—what a mess! Where do I start?" As we move along, you'll see some tricks for how to make your programs more readable and easier to maintain.

The Code, Step by Step

At first, you might consider this kind of explanation too unnecessary, a bit like when I was in school and I had to study Dante's *Divina Commedia* (every Italian student has to go through that, as well as another book called *I promessi sposi*, or *The Betrothed*—oh, the nightmares). For each line of the poems, there were a hundred lines of commentary! However, the explanation will be much more useful here as you move on to writing your own programs.

—Massimo

```
// Blinking LED
```

A comment is a useful way for us to write little notes. The preceding title comment just reminds us that this program, Example 4-1, blinks an LED.

```
const int LED = 13; // LED connected to
                    // digital pin 13
```

const int means that *LED* is the name of an integer number that can't be changed (i.e., a constant) whose value is set to 13. It's like an automatic search-and-replace for your code; in this case, it's telling Arduino to write the number 13 every time the word *LED* appears.

The reason we need the number 13 is that the preinstalled LED we mentioned earlier is attached to Arduino pin 13. A common convention is to use uppercase letters for constants.

```
void setup()
```

This line tells Arduino that the next block of code will be a function named **setup()**.

```
{
```

With this opening curly brace, a block of code begins.

```
pinMode(LED, OUTPUT); // sets the digital
                      // pin as output
```

Finally, a really interesting instruction! **pinMode()** tells Arduino how to configure a certain pin. Digital pins can be used either as input or output, but we need to tell Arduino how we intend to use the pin.

In this case, we need an output pin to control our LED.

pinMode() is a function, and the words (or numbers) specified inside the parentheses are called its *arguments*. Arguments are whatever information a function needs in order to do its job.

The **pinMode()** function needs two arguments. The first argument tells **pinMode()** which pin we're talking about, and the second argument tells **pinMode()** whether we want to use that pin as an input or output. **INPUT** and **OUTPUT** are predefined constants in the Arduino language.

Remember that the word *LED* is the name of the constant which was set to the number 13, which is the pin number to which the LED is attached. So, the first argument is `LED`, the name of the constant.

The second argument is `OUTPUT`, because when Arduino talks to an actuator, it's sending information *out*.

```
}
```

This closing curly brace signifies the end of the `setup()` function.

```
void loop()
{
```

`loop()` is where you specify the main behaviour of your interactive device. It will be repeated over and over again until you remove power from the board.

```
digitalWrite(LED, HIGH);   // turns the LED on
```

As the comment says, `digitalWrite()` is able to turn on (or off) any pin that has been configured as an output. Just as we saw with the `pinMode()` function, `digitalWrite()` expects two arguments, and just as we saw with the `pinMode()` function, the first argument tells `digitalWrite()` what pin we're talking about, and just as we saw with the `pinMode()` function, we'll use the constant name `LED` to refer to pin number 13, which is where the preinstalled LED is attached.

The second argument is different: in this case, the second argument tells `digitalWrite()` whether to set the voltage level to 0 (`LOW`) or to 5 V (`HIGH`).

Imagine that every output pin is a tiny power socket, like the ones you have on the walls of your apartment. European ones are 230 V, American ones are 110 V, and Arduino works at a more modest 5 V. The magic here is when software can control hardware. When you write `digitalWrite(LED, HIGH)`, it turns the output pin to 5 V, and if you connect an LED, it will light up. So at this point in your code, an instruction in software makes something happen in the physical world by controlling the flow of electricity to the pin. Turning on and off the pin will now let us translate these into something more visible for a human being; the LED is our *actuator*.

On the Arduino, HIGH means that the pin will be set to 5 V, while LOW means the pin will be set to 0 V.

You might wonder why we use HIGH and LOW instead of ON and OFF. It's true that HIGH or LOW usually correspond to on and off, respectively, but this depends on how the pin is used. For example, an LED connected between 5V and a pin will turn on when that pin is LOW and turn off when the pin is HIGH. But for most cases you can just pretend that HIGH means ON and LOW means OFF.

```
delay(1000);      // waits for a second
```

Although Arduino is much slower than your laptop, it's still very fast. If we turned the LED on and then immediately turned it off, our eyes wouldn't be able to see it. We need to keep the LED on for a while so that we can see it, and the way to do that is to tell Arduino to wait for a while before going to the next step. `delay()` basically makes the microcontroller sit there and do nothing for the amount of milliseconds that you pass as an argument. Milliseconds are thousandths of seconds; therefore, 1,000 milliseconds equals 1 second. So the LED stays on for 1 second here.

```
digitalWrite(LED, LOW);      // turns the LED off
```

This instruction now turns off the LED that we previously turned on.

```
delay(1000); // waits for a second
```

Here, we delay for another second. The LED will be off for 1 second.

```
}
```

This closing curly brace marks the end of the `loop()` function. When Arduino gets to this, it starts over again at the beginning of `loop()`.

To sum up, this program does this:

- Turns pin 13 into an output (just once at the beginning)
- Enters a loop
- Switches on the LED connected to pin 13

- Waits for a second
- Switches off the LED connected to pin 13
- Waits for a second
- Goes back to beginning of the loop

We hope that wasn't too painful. If you didn't understand everything, don't feel discouraged. As we mentioned before, if you're new to these concepts, it takes a while before they make sense. You'll learn more about programming as you go through the later examples.

Before we move on to the next section, we want you to play with the code. For example, reduce the amount of delay, using different numbers for the on and off pulses so that you can see different blinking patterns. In particular, you should see what happens when you make the delays very small, but use different delays for on and off. There is a moment when something strange happens; this "something" will be very useful when you learn about *pulse-width modulation* in "Controlling Light with PWM" on page 56.

What We Will Be Building

> I have always been fascinated by light and the ability to control different light sources through technology. I have been lucky enough to work on some interesting projects that involve controlling light and making it interact with people. Arduino is really good at this.
>
> —Massimo

In this chapter, Chapter 5, and Chapter 7, we will be working on how to design *interactive lamps*, using Arduino as a way to learn the basics of how interactive devices are built. Remember, though, that Arduino doesn't really understand, or care, what you connect to the output pins. Arduino just turns the pin HIGH or LOW, which could be controlling a light, or an electric motor, or your car engine.

In the next section, we'll explain the basics of electricity in a way that would bore an engineer, but won't scare a new Arduino programmer.

What Is Electricity?

If you have done any plumbing at home, electronics won't be a problem for you to understand. To understand how electricity and electric circuits work, the best way is to use something called the *water analogy*. Let's take a simple device, like the battery-powered portable fan shown in Figure 4-3.

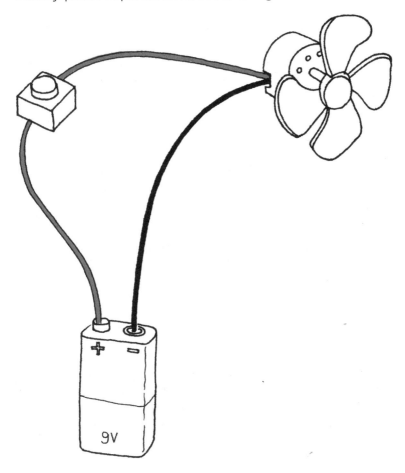

Figure 4-3. *A portable fan*

If you take a fan apart, you will see that it contains a battery, a couple of wires, and an electric motor, and that one of the wires going to the motor is interrupted by a switch. If you turn the switch on, the motor will start to spin, providing the necessary airflow to cool you down.

How does this work? Well, imagine that the battery is both a water reservoir and a pump, the switch is a tap, and the motor is one of those wheels that you see in watermills. When you open the tap, water flows from the pump and pushes the wheel into motion.

In this simple hydraulic system, shown in Figure 4-4, two factors are important: the pressure of the water (this is determined by the power of the pump) and the amount of water that will flow in the pipes (this depends on the size of the pipes and the *resistance* that the wheel will provide to the stream of water hitting it).

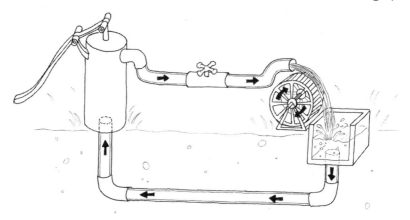

Figure 4-4. *A hydraulic system*

You'll quickly realise that if you want the wheel to spin faster, you need to increase the size of the pipes (but this works only up to a point) and increase the pressure that the pump can achieve. Increasing the size of the pipes allows a greater flow of water to go through them; by making them bigger, you have effectively reduced the pipes' resistance to the flow of water. This approach works up to a certain point, at which the wheel won't spin any faster, because the pressure of the water is not strong enough. When you reach this point, you need the pump to be stronger. This method of speeding up the watermill can go on until the point when the wheel falls apart because the water flow is too strong for it and it is destroyed. Another thing you will notice is that as the wheel spins, the axle will heat up a little bit, because no matter how well you have mounted the wheel, the friction between the axle and the holes in which it is mounted will generate heat. It is important to understand that in a system like this, not all the energy you pump into the system will be converted into movement; some will be lost in a number of inefficiencies and will generally show up as heat emanating from some parts of the system.

So what are the important parts of the system? The pressure produced by the pump is one; the resistance that the pipes and wheel offer to the flow of water, and the actual flow of water (let's say that this is represented by the number of litres of water that flow in one second) are the others.

Electricity works a bit like water. You have a kind of pump (any source of electricity, like a battery or a wall plug) that pushes electric charges (imagine them as "drops" of electricity) down pipes, which are represented by the wires. Various electrical devices are able to use these drops of electricity to produce heat (your grandma's electric blanket), light (your bedroom lamp), sound (your stereo), movement (your fan), and much more.

When you read that a battery's voltage is 9 V, think of this voltage as the water pressure that can potentially be produced by this little "pump". *Voltage* is measured in volts, named after Alessandro Volta, the inventor of the first battery.

Just as water pressure has an electric equivalent, the flow rate of water does too. This is called *current*, and is measured in amperes (after André-Marie Ampère, electromagnetism pioneer). The relationship between voltage and current can be illustrated by returning to the water wheel: a higher voltage (pressure) lets you spin a wheel faster; a higher flow rate (current) lets you spin a larger wheel.

Finally, the resistance opposing the flow of current over any path that it travels is called—you guessed it—*resistance*, and is measured in ohms (after the German physicist Georg Ohm).

Herr Ohm was also responsible for formulating the most important law in electricity—and the only formula that you really need
to remember. He was able to demonstrate that in a circuit, the
voltage, the current, and the resistance are all related to each
other, and in particular that the resistance of a circuit deter-
mines the amount of current that will flow through it, given a
certain voltage supply.

It's very intuitive, if you think about it. Take a 9 V battery and
plug it into a simple circuit. While measuring current, you will
find that the more resistors you add to the circuit, the less current will travel through it. Going back to the analogy of water
flowing in pipes, given a certain pump, if I install a valve (which
we can relate to a variable resistor in electricity), the more I
close the valve—increasing resistance to water flow—the less
water will flow through the pipes. Ohm summarised his law in
these formulas:

```
R (resistance) = V (voltage) / I (current)
V = R * I
I = V / R
```

What's important about this law is understanding it intuitively,
and for this, I prefer the last version (I = V / R) because the
current is something that results when you apply a certain voltage (the pressure) to a certain circuit (the resistance). The voltage exists whether or not it's being used, and the resistance
exists whether or not it's being given electricity, but the current
comes into existence only when these are put together.

Using a Pushbutton to Control the LED

Blinking an LED was easy, but I don't think you would stay sane
if your desk lamp were to continuously blink while you were trying to read a book. Therefore, you need to learn how to control
it. In the previous example, the LED was your actuator, and the
Arduino was controlling it. What's missing to complete the picture is a sensor.

In this case, we're going to use the simplest form of sensor available: a pushbutton switch.

If you were to take apart a pushbutton, you would see that it is a very simple device: two bits of metal kept apart by a spring, and a plastic cap that when pressed brings the two bits of metal into contact. When the bits of metal are apart, there is no circulation of current in the pushbutton (a bit like when a water valve is closed); when you press it, you make a connection.

All switches are basically just this: two (or more) pieces of metal that can be brought into contact with each other, allowing electricity to flow from one to the other, or separated, preventing the flow of electricity.

To monitor the state of a switch, there's a new Arduino instruction that you're going to learn: the `digitalRead()` function.

`digitalRead()` checks to see whether there is any voltage applied to the pin that you specify between parentheses, and returns a value of HIGH or LOW, depending on its findings. The other instructions that you've used so far haven't returned any information—they just executed what we asked them to do. But that kind of function is a bit limited, because it will force you to stick with very predictable sequences of instructions, with no input from the outside world. With `digitalRead()`, you can "ask a question" of Arduino and receive an answer that can be stored in memory somewhere and used to make decisions immediately or later.

Build the circuit shown in Figure 4-5. To build this, you'll need to obtain some parts (these will come in handy as you work on other projects as well):

Solderless breadboard
 Maker Shed part number MKKN3, Arduino Store (*http://arduino.cc/breadboard*). Appendix A is an introduction to the solderless breadboard.

Precut jumper wire kit
 Maker Shed MKSEEED3, Arduino Store (included with the breadboard just listed).

One 10 K ohm resistor
 Arduino Store (*http://arduino.cc/resistor10k*), 10-pack.

Momentary tactile pushbutton switch
    Arduino Store (*http://arduino.cc/pushButton*), 10-pack.

 Instead of buying precut jumper wire, you can also buy 22 AWG solid-core hookup wire in small spools (such as Maker Shed MKEE3) and then cut and strip it yourself using wire cutters and wire strippers.

 GND on the Arduino board stands for *ground*. The word is historical, but in our case simply means the negative side of the power. We tend to use the words GND and ground interchangeably. You can think of it as the pipe that's underground in the water analogy in Figure 4-4.

In most circuits, GND or ground is used very frequently. For this reason, your Arduino board has three pins labeled GND. They are all connected together, and it makes no difference which one you use.

The pin labeled *5V* is the positive side of the power, and is always 5 volts higher than the ground.

Example 4-2 shows the code that we'll be using to control the LED with our pushbutton.

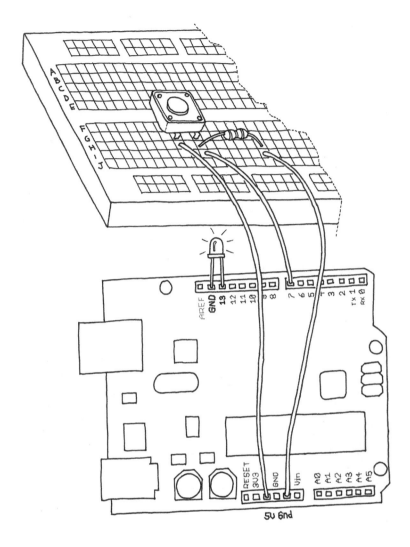

Figure 4-5. *Hooking up a pushbutton*

Example 4-2. Turn on LED while the button is pressed

```
// Turn on LED while the button is pressed

const int LED = 13;   // the pin for the LED
```

```
const int BUTTON = 7;  // the input pin where the
                       // pushbutton is connected
int val = 0;           // val will be used to store the state
                       // of the input pin

void setup() {
  pinMode(LED, OUTPUT);   // tell Arduino LED is an output
  pinMode(BUTTON, INPUT); // and BUTTON is an input
}

void loop(){
  val = digitalRead(BUTTON); // read input value and store it

  // check whether the input is HIGH (button pressed)
  if (val == HIGH) {
    digitalWrite(LED, HIGH); // turn LED ON
  } else {
    digitalWrite(LED, LOW);
  }
}
```

In Arduino, select File→New (if you have another sketch open,
you may want to save it first). When Arduino asks you to name
your new sketch folder, type *PushButtonControl*. Type the
Example 4-2 code into Arduino (or download it from this book's
catalog page (*http://bit.ly/start_arduino_3e*) and paste it into
the Arduino IDE). If everything is correct, the LED will light up
when you press the button.

How Does This Work?

We have introduced two new concepts with this example program: functions that return the result of their work, and the if
statement.

The if statement is possibly the most important instruction in a
programming language, because it allows a computer (and
remember, the Arduino is a small computer) to make decisions.
After the if keyword, you have to write a "question" inside
parentheses, and if the "answer", or result, is true, the first block
of code will be executed; otherwise, the block of code after else
will be executed.

Notice that the == symbol is very different from the = symbol.
The former is used when two entities are compared, and returns

`true` or `false`; the latter assigns a value to a constant or a variable. Make sure that you use the correct one, because it is very easy to make that mistake and use just =, in which case your program will never work. We know, because after years of programming, we still make that mistake.

It's important to realise that the switch is not connected directly to the LED. Your Arduino sketch inspects the switch, and then makes a decision as to whether to turn the LED on or off. The connection between the switch and the LED is really happening in your sketch.

Holding your finger on the button for as long as you need light is not practical. Although it would make you think about how much energy you're wasting when you walk away from a lamp that you left on, we need to figure out how to make the on button "stick".

One Circuit, a Thousand Behaviours

The great advantage of digital, programmable electronics over classic electronics now becomes evident: I will show you how to implement many different "behaviours" using the same electronic circuit as in the previous section, just by changing the software.

As I've mentioned before, it's not very practical to have to hold your finger on the button to have the light on. You therefore must implement some form of "memory", in the form of a software mechanism that will remember when you have pressed the button and will keep the light on even after you have released it.

To do this, you're going to use what is called a *variable*. (You have used one already, but we haven't explained it.) A variable is a place in the Arduino memory where you can store data. Think of it like one of those sticky notes you use to remind yourself about something, such as a phone number: you take one, you write "Luisa 02 555 1212" on it, and you stick it to your computer monitor or your fridge. In the Arduino language, it's equally simple: you just decide what type of data you want to store (a number or some text, for example), give it a name, and when you want to, you can store the data or retrieve it. For example:

```
int val = 0;
```

int means that your variable will store an integer number, val is the name of the variable, and = 0 assigns it an initial value of zero.

A variable, as the name intimates, can be modified anywhere in your code, so that later on in your program, you could write:

```
val = 112;
```

which reassigns a new value, 112, to your variable.

 Have you noticed that in Arduino, every instruction ends with a semicolon? This is done so the compiler (the part of Arduino that turns your sketch into a program that the microcontroller can run) knows your statement is finished and a new one is beginning. If you forget a semicolon where one is required, the compiler won't be able to make sense of your sketch.

In the following program, the variable val stores the result of digitalRead(); whatever Arduino gets from the input ends up in the variable and will stay there until another line of code changes it. Notice that variables use a type of memory called *RAM*. It is quite fast, but when you turn off your board, all data stored in RAM is lost (which means that each variable is reset to its initial value when the board is powered up again). Your programs themselves are stored in flash memory—this is the same type used by your mobile phone to store phone numbers—which retains its content even when the board is off.

Let's now use another variable to remember whether the LED has to stay on or off after we release the button. Example 4-3 is a first attempt at achieving that.

Example 4-3. Turn on LED when the button is pressed and keep it on after it is released

```
const int LED = 13;   // the pin for the LED
const int BUTTON = 7; // the input pin where the
                      // pushbutton is connected
int val = 0;    // val will be used to store the state
                // of the input pin
int state = 0;  // 0 = LED off while 1 = LED on

void setup() {
  pinMode(LED, OUTPUT);    // tell Arduino LED is an output
  pinMode(BUTTON, INPUT); // and BUTTON is an input
}

void loop() {
  val = digitalRead(BUTTON); // read input value and store it

  // check if the input is HIGH (button pressed)
  // and change the state
  if (val == HIGH) {
    state = 1 - state;
  }

  if (state == 1) {
    digitalWrite(LED, HIGH); // turn LED ON
  } else {
    digitalWrite(LED, LOW);
  }
}
```

Now go test this code. You will notice that it works...somewhat. You'll find that the light changes so rapidly that you can't reliably set it on or off with a button press.

Let's look at the interesting parts of the code: state is a variable that stores either 0 or 1 to remember whether the LED is on or off. After the button is released, we initialise it to 0 (LED off).

Later, we read the current state of the button, and if it's pressed (val == HIGH), we change state from 0 to 1, or vice versa. We do this using a small trick, as state can be only either 1 or 0. The trick I use involves a small mathematical expression based on the idea that 1 – 0 is 1 and 1 – 1 is 0:

```
state = 1 - state;
```

The line may not make much sense in mathematics, but it does in programming. The symbol = means "assign the result of what's after me to the variable name before me"—in this case, the new value of state is assigned the value of 1 minus the old value of state.

Later in the program, you can see that we use state to figure out whether the LED has to be on or off. As I mentioned, this leads to somewhat flaky results.

The results are flaky because of the way we read the button. Arduino is really fast; it executes its own internal instructions at a rate of 16 million per second—it could well be executing a few million lines of code per second. So this means that while your finger is pressing the button, Arduino might be reading the button's position a few thousand times and changing state accordingly. So the results end up being unpredictable; it might be off when you wanted it on, or vice versa. As even a broken clock is right twice a day, the program might show the correct behaviour every once in a while, but much of the time it will be wrong.

How do you fix this? Well, you need to detect the exact moment when the button is pressed—that is the only moment that you have to change state. The way we like to do it is to store the value of val before we read a new one; this allows you to compare the current position of the button with the previous one and change state only when the button changes from LOW to HIGH.

Example 4-4 contains the code to do so.

Example 4-4. New and improved button press formula!

```
const int LED = 13;   // the pin for the LED
const int BUTTON = 7; // the input pin where the
                      // pushbutton is connected
int val = 0;       // val will be used to store the state
                   // of the input pin
int old_val = 0; // this variable stores the previous
                 // value of "val"
```

```
int state = 0;    // 0 = LED off and 1 = LED on

void setup() {
  pinMode(LED, OUTPUT);    // tell Arduino LED is an output
  pinMode(BUTTON, INPUT); // and BUTTON is an input
}
void loop(){
  val = digitalRead(BUTTON); // read input value and store it
                             // yum, fresh

  // check if there was a transition
  if ((val == HIGH) && (old_val == LOW)){
    state = 1 - state;
  }

  old_val = val;  // val is now old, let's store it

  if (state == 1) {
    digitalWrite(LED, HIGH); // turn LED ON
  } else {
    digitalWrite(LED, LOW);
  }
}
```

Test it: you're almost there!

You may have noticed that this approach is not entirely perfect, due to another issue with mechanical switches.

As we explained earlier, pushbuttons are just two bits of metal kept apart by a spring, that come into contact when you press the button. This might seem like the switch should be completely on when you press the button, but in fact what happens is the two pieces of metal bounce off each other, just like a ball bounces on the floor.

Although the bouncing is only for a very small distance and happens for a fraction of a second, it causes the switch to change between off and on a number of times until the bouncing stops, and Arduino is quick enough to catch this.

When the pushbutton is bouncing, the Arduino sees a very rapid sequence of on and off signals. There are many techniques developed to do *debouncing*, but in this simple piece of code, it's usually enough to add a 10- to 50-millisecond delay when the code detects a transition. In other words, you just wait a bit for the bouncing to stop.

Example 4-5 is the final code.

Example 4-5. Another new and improved formula for button presses—with simple debouncing!

```
const int LED = 13;     // the pin for the LED
const int BUTTON = 7;   // the input pin where the
                        // pushbutton is connected
int val = 0;       // val will be used to store the state
                   // of the input pin
int old_val = 0; // this variable stores the previous
                 // value of "val"
int state = 0;    // 0 = LED off and 1 = LED on

void setup() {
  pinMode(LED, OUTPUT);   // tell Arduino LED is an output
  pinMode(BUTTON, INPUT); // and BUTTON is an input
}

void loop(){
 val = digitalRead(BUTTON); // read input value and store it
                            // yum, fresh

 // check if there was a transition
 if ((val == HIGH) && (old_val == LOW)){
   state = 1 - state;
   delay(10);
 }

 old_val = val; // val is now old, let's store it

 if (state == 1) {
   digitalWrite(LED, HIGH); // turn LED ON
 } else {
   digitalWrite(LED, LOW);
 }
}
```

One reader, Tami (Masaaki) Takamiya, wrote in with some extra code that may give you better debouncing:

```
if ((val == LOW) && (old_val == HIGH)) {
    delay(10);
}
```

5/Advanced Input and Output

What you have just learned in Chapter 4 are the most elementary operations we can do in Arduino: controlling digital output and reading digital input. If Arduino were some sort of human language, those would be two letters of its alphabet. Considering that there are just five letters in this alphabet, you can see how much more work we have to do before we can write Arduino poetry.

Trying Out Other On/Off Sensors

Now that you've learned how to use a pushbutton, you should know that there are many other very basic sensors that work according to the same principle:

Toggle switch
> The pushbutton that you've been using is a type of switch called a *momentary* switch, because once you let it go, it goes back to where it was. A common example of a momentary switch is a doorbell.

> In contrast, a *toggle* switch stays where you put it. A common example of a toggle switch is a light switch.

> In spite of these technically correct definitions, in this book we'll use the common names for these switches: a pushbutton refers to a momentary switch, while a switch refers to a toggle switch.

> Although you might not think of a switch as a sensor, in fact it is: a pushbutton (momentary switch) senses when you are pressing it, and a (toggle) switch senses and remembers the last state you put it in.

Thermostat

A switch that changes state when the temperature reaches a set value.

Magnetic switch (also known as a reed switch)

Has two contacts that come together when they are near a magnet; often used for burglar alarms to detect when a door or window is opened.

Carpet switch

Mat that can be placed under a carpet or a doormat to detect when a human being (or heavy cat) steps on them.

Tilt switch or tilt sensor

A simple but clever sensor that contains two (or more) contacts and a little metal ball (or a drop of mercury, but I don't recommend using that). Figure 5-1 shows the inside of a typical model.

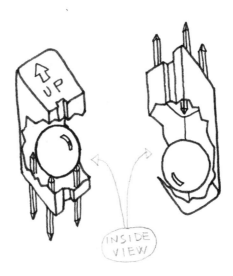

Figure 5-1. *The inside of a tilt sensor*

When the sensor is in its upright position, the ball bridges the two contacts, and this works just as if you had pressed a pushbutton. When you tilt this sensor, the ball moves, and the contact is opened, which is just as if you had released a pushbutton. Using this simple component, you can implement, for example, gestural interfaces that react when an object is moved or shaken.

Another useful sensor that is often found in burglar alarms is the *passive infrared*, or *PIR*, sensor, shown in Figure 5-2. This device changes state when a human moves within its proximity. These sensors are often designed to detect humans but not animals, so that burglar alarms won't get triggered by pets.

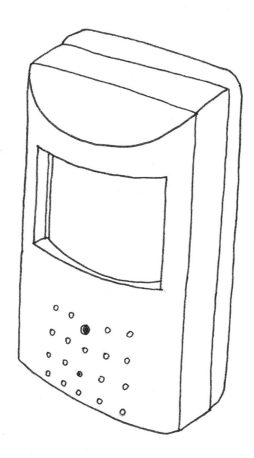

Figure 5-2. *Typical PIR sensor*

Detecting humans is actually quite complicated, and the PIR sensor is quite complex inside. Fortunately, we don't really care about the insides. The only thing we need to know is that it results in a digital signal, indicating whether a human is present or not. This is why the PIR sensor is a digital sensor.

Homemade (DIY) Switches

You can make your own tilt switch with a metal ball and a few nails, and by wrapping some wire around the nails. When the ball rolls to one side and rests on two of the nails, it will connect those wires.

You can make a momentary switch with a clothespin by wrapping a wire around each handle. When the handles are squeezed, the switch is closed.

Alternatively, you can wrap the wires around the jaws of the clothespin, and put a piece of cardboard between them to keep the wires from touching. Tie a piece of string to the cardboard and tie the other end of the string to a door. When the door is opened, the string pulls the cardboard out, the wires touch, and the switch is closed.

Because all of these sensors are digital, any of them can be used in place of the pushbutton that you worked with in Chapter 4 without having to make any changes to your sketch.

For example, by using the circuit and sketch from "Using a Pushbutton to Control the LED" on page 40, but replacing the pushbutton with a PIR sensor, you could make your lamp respond to the presence of human beings, or you could use a tilt switch to build one that turns off when it's tilted on one side.

Controlling Light with PWM

You already know enough to build an interactive lamp, but so far the result is a little boring, because the light is only either on or off. A fancy interactive lamp needs to be dimmable. To solve this problem, we can use a little trick that makes a lot of things possible, such as TV or cinema. This trick is called *Persistence of Vision*, or *POV*.

As we hinted at after the first example in Chapter 4, if you reduce the numbers in the delay function until you don't see the LED blinking anymore, you will notice that the LED seems to be dimmer than its normal brightness. If you experiment with this, you will notice that if you make the on delay different from the off delay, you can make the LED brighter by leaving it on for longer, and you can make the LED dimmer by leaving it off for longer. This technique is called *pulse-width modulation*, or *PWM*, because you are changing the LED's brightness by modulating (or changing) the width of the pulse. Figure 5-3 shows how this works. This works because our eyes can't see distinct pictures if they change too fast.

This technique works with some devices other than an LED. For example, you can change the speed of a motor in the same way. In this case, it isn't our eyes that allow this to happen, but rather the motor itself, because it can't start or stop turning instantly. It takes a small amount of time to speed up and slow down. If we change the output (using digitalWrite()) faster than the motor can respond, it ends up turning at some intermediate speed, depending on how much time it's turned on, and how much time it's turned off.

While this trick is very useful, you probably felt that controlling the LED brightness by fiddling with the delays in your code is a bit inconvenient. Even worse, as soon as you want to read a sensor, send data on the serial port, or do almost anything else, the LED brightness will change, because any extra lines of code you add will take time to execute, which will change the amount of time the LED is on or off.

Luckily, the microcontroller used by your Arduino has a piece of hardware that can very efficiently blink your LEDs while your sketch does something else. On the Uno, this hardware is implemented on pins 3, 5, 6, 9, 10, and 11, and on the Leonardo, on pins 3, 5, 6, 9, 10, 11, and 13. The analogWrite() instruction is used to control this hardware.

PWM

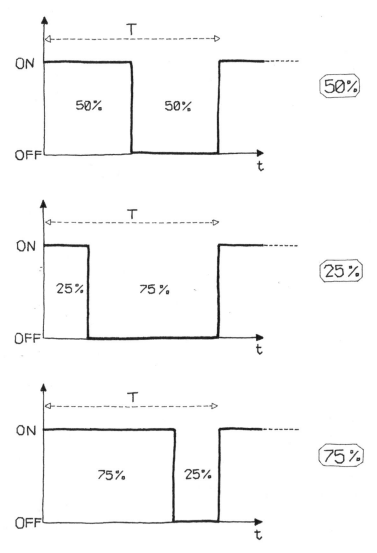

Figure 5-3. *PWM in action*

For example, writing `analogWrite(9,50)` will set the brightness of an LED connected to pin 9 to quite dim, while writing **analog-**

`Write(9,200)` will set the brightness of the LED to quite bright. `analogWrite()` takes a number between 0 and 255, where 255 means full brightness and 0 means off.

 Having multiple PWM pins is very useful. For example, if you buy red, green, and blue LEDs, you can mix their lights and make light of any colour.

Let's try it out. Build the circuit that you see in Figure 5-4. You'll need an LED of any colour (e.g., *a green version*) and some 220-ohm resistors (e.g., *available in the Arduino store*).

Note that LEDs are *polarized*, which means they care which way the electric current goes through them. The long lead indicates the *anode*, or positive lead, and in our case should go to the right, which connects it to pin 9 of the Arduino. The short lead indicates the *cathode*, or negative lead, and in this case it should go to the left, connecting it to the resistor.

Most LEDs also have a flattened edge on the cathode side, as shown in the figure. An easy way to remember this is that the flat part looks like a minus sign.

As I mentioned in "Blinking an LED" on page 26, you should always use a resistor with an LED to prevent burning out the LED. Use a 220-ohm resistor (red-red-brown).

Then, create a new sketch in Arduino with the code shown in Example 5-1. You can also download it from the example code link on the book's catalog page (*http://bit.ly/start_arduino_3e*).

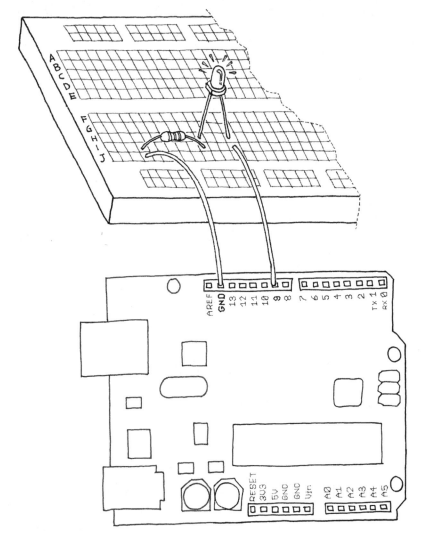

Figure 5-4. *LED connected to PWM pin*

Example 5-1. Fade an LED in and out, like on a sleeping Apple computer

```
const int LED = 9; // the pin for the LED
int i = 0;         // We'll use this to count up and down
```

```
void setup() {
  pinMode(LED, OUTPUT); // tell Arduino LED is an output
}

void loop(){

  for (i = 0; i < 255; i++) { // loop from 0 to 254 (fade in)
    analogWrite(LED, i);      // set the LED brightness
    delay(10); // Wait 10ms because analogWrite
               // is instantaneous and we would
               // not see any change
  }

  for (i = 255; i > 0; i--) { // loop from 255 to 1 (fade out)

    analogWrite(LED, i); // set the LED brightness
    delay(10);           // Wait 10ms
  }

}
```

Upload the sketch, and the LED will fade up and then fade down continuously. Congratulations! You have replicated a fancy feature of a laptop computer.

Maybe it's a bit of a waste to use Arduino for something so simple, but you can learn a lot from this example.

As you learned earlier, analogWrite() changes the LED brightness. The other important part is the for loop: it repeats the analogWrite() and the delay() over and over, each time using a different value for the variable i as follows.

The first for loop starts the variable i with the value of 0, and increases it up to 255, which fades the LED up to full brightness.

The second for loop starts the variable i with the value of 255, and decreases it up to 0, which fades the LED all the way down to completely off.

After the second for loop, Arduino starts our loop() function over again.

The delay() is just to slow things down a bit so you can see the changing brightness; otherwise, it would happen too fast.

Let's use this knowledge to improve our lamp.

Add the circuit we used to read a button (back in Chapter 4) to this breadboard. See if you can do this without reading past this paragraph, because I want you to start thinking about the fact that each elementary circuit I show here is a building block to make bigger and bigger projects. If you need to peek ahead, don't worry; the most important thing is that you spend some time thinking about how it might look.

To create this circuit, you will need to combine the circuit you just built (shown in Figure 5-4) with the pushbutton circuit shown in Figure 4-5. If you'd like, you can simply build both circuits on different parts of the breadboard; you have plenty of room.

Take a look at Appendix A to learn more about the solderless breadboard.

If you're not ready to try this, don't worry: simply wire up both circuits to your Arduino as shown in Figure 4-5 and Figure 5-4.

Getting back to this next example, if you have just one pushbutton, how do you control the brightness of a lamp? You're going to learn yet another Interaction Design technique: detecting how long a button has been pressed. To do this, we need to upgrade Example 4-5 from Chapter 4 to add dimming. The idea is to build an interface in which a press-and-release action switches the light on and off, and a press-and-hold action changes brightness.

Have a look at the sketch in Example 5-2. It turns on the LED when the button is pressed and keeps it on after it is released. If the button is held, the brightness changes.

Example 5-2. Sketch to change the brightness as you hold the button

```
const int LED = 9;      // the pin for the LED
const int BUTTON = 7;   // input pin of the pushbutton

int val = 0;        // stores the state of the input pin

int old_val = 0; // stores the previous value of "val"
int state = 0;     // 0 = LED off while 1 = LED on
```

```
int brightness = 128;        // Stores the brightness value
unsigned long startTime = 0; // when did we begin pressing?

void setup() {
  pinMode(LED, OUTPUT);    // tell Arduino LED is an output
  pinMode(BUTTON, INPUT);  // and BUTTON is an input
}

void loop() {

  val = digitalRead(BUTTON); // read input value and store it
                             // yum, fresh

  // check if there was a transition
  if ((val == HIGH) && (old_val == LOW)) {

    state = 1 - state; // change the state from off to on
                       // or vice-versa

    startTime = millis(); // millis() is the Arduino clock
                          // it returns how many milliseconds
                          // have passed since the board has
                          // been reset.

    // (this line remembers when the button
    // was last pressed)
    delay(10);
  }

  // check whether the button is being held down
    if ((val == HIGH) && (old_val == HIGH)) {

      // If the button is held for more than 500 ms.
      if (state == 1 && (millis() - startTime) > 500) {

        brightness++; // increment brightness by 1
        delay(10);    // delay to avoid brightness going
                      // up too fast

        if (brightness > 255) { // 255 is the max brightness

          brightness = 0; // if we go over 255
                          // let's go back to 0
        }
      }
```

```
  }

  old_val = val; // val is now old, let's store it

  if (state == 1) {
    analogWrite(LED, brightness); // turn LED ON at the
                                  // current brightness level
  } else {
    analogWrite(LED, 0); // turn LED OFF
  }
}
```

Now try it out. As you can see, this interaction model is taking shape. If you press the button and release it immediately, you switch the lamp on or off. If you hold the button down, the brightness changes; just let go when you have reached the desired brightness.

Just as we said before about thinking about the circuit, try to spend a bit of time trying to understand the program.

Probably the most confusing line is this one:

```
    if (state == 1 && (millis() - startTime) > 500) {
```

This checks to see if the button is held down for more than 500 ms by using a built-in function called millis(), which is just a running count of the number of milliseconds since your sketch started running. By keeping track of when the button was pressed (in the variable startTime), we can compare the current time to the start time to see how much time has passed.

Of course, this makes sense only if the button is currently pressed, which is why at the beginning of the line we check to see if state is set to the value of 1.

As you can see, switches are really pretty powerful sensors, even though they are so simple. Now let's learn how to use some other sensors.

Use a Light Sensor Instead of the Pushbutton

Now we're going to try an interesting experiment using a light sensor, like the one pictured in Figure 5-5. You can get these from electronics suppliers such as RadioShack, as part of the Maker Shed Getting Started with Arduino Kit (*http://bit.ly/get-started-arduino-v3*), or from the Arduino store (*http://bit.ly/ArduinoStoreLDR*).

Figure 5-5. *Light-dependent resistor (LDR)*

As its name suggests, the *light-dependent resistor* (*LDR*) is some sort of resistor that depends on light. In darkness, the resistance of an LDR is quite high, but when you shine some light at it, the resistance quickly drops and it becomes a reasonably good conductor of electricity. It is thus a kind of light-activated switch.

Build the circuit shown in Figure 4-5 (see "Using a Pushbutton to Control the LED" on page 40 in Chapter 4), and then upload the code from Example 4-2 to your Arduino. Press the pushbutton to make sure it works.

Now carefully remove only the pushbutton, and insert the LDR into the circuit exactly where the pushbutton was. The LED should come on. Cover the LDR with your hands, and the LED turns off.

You've just built your first real sensor-driven LED. This is important because for the first time in this book, you are using an electronic component that is not a simple mechanical device: it's a real, rich sensor. In fact, this is only a small example of what the LDR can be used for.

Analogue Input

As you learned in the previous section, Arduino is able to detect whether there is a voltage applied to one of its pins and report it through the `digitalRead()` function. This kind of either/or response is fine in a lot of applications, but the light sensor that we just used is able to tell us not only whether there is light, but also how much light there is. This is the difference between an on/off or *digital* sensor (which tells us whether something is there or not) and an *analogue* sensor, which can tell us how much of something there is.

In order to read this type of sensor, we need to use a special Arduino pin.

Turn your Arduino around so it matches Figure 5-6.

In the top-left part of the board, you'll see six pins marked Analog In; these are special pins that not only can tell you whether there is a voltage applied to them, but also can measure the amount of that voltage, by using the `analogRead()` function. The `analogRead()` function returns a number between 0 and 1023, which represents voltages between 0 and 5 volts. For example, if there is a voltage of 2.5 V applied to pin number 0, `analogRead(0)` returns 512.

If you now build the circuit that you see in Figure 5-6, using a 10 K ohm resistor, and run the code listed in Example 5-3, you'll see the onboard LED blinking at a rate that depends on the amount of light that shines on the sensor.

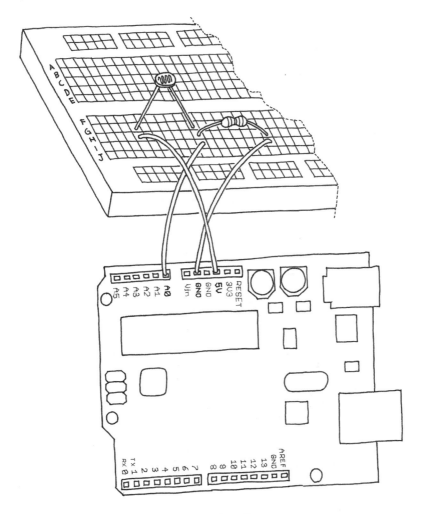

Figure 5-6. *An analogue sensor circuit*

Example 5-3. Blink LED at a rate specified by the value of the analogue input

```
const int LED = 13; // the pin for the LED

int val = 0;    // variable used to store the value
                // coming from the sensor
void setup() {
  pinMode(LED, OUTPUT); // LED is as an OUTPUT
```

```
  // Note: Analogue pins are
  // automatically set as inputs
}

void loop() {

  val = analogRead(0); // read the value from
                       // the sensor

  digitalWrite(LED, HIGH); // turn the LED on

  delay(val); // stop the program for
              // some time

  digitalWrite(LED, LOW); // turn the LED off

  delay(val); // stop the program for
              // some time
}
```

Now add an LED to pin 9 as we did before, using the circuit shown in Figure 5-4. Because you already have some stuff on the breadboard, you'll need to find a spot on the breadboard where the LED, wires, and resistor won't overlap with the LDR circuit. You may have to move some things around, but don't worry, this is good practice because it helps your understanding of circuits and the breadboard.

When you are done adding the LED to your LDR circuit, type in Example 5-4 and upload it to your Arduino.

Example 5-4. Set the LED to a brightness specified by the value of the analogue input

```
const int LED = 9;  // the pin for the LED

int val = 0;   // variable used to store the value
               // coming from the sensor

void setup() {

  pinMode(LED, OUTPUT); // LED is as an OUTPUT

  // Note: Analogue pins are
```

```
    // automatically set as inputs
}

void loop() {

    val = analogRead(0); // read the value from
                         // the sensor
    analogWrite(LED, val/4); // turn the LED on at
                             // the brightness set
                             // by the sensor

    delay(10); // stop the program for
               // some time
}
```

Once it's running, cover and uncover the LDR and see what happens to the LED brightness.

As before, try to understand what's going on. The program is really very simple, in fact much simpler than the previous two examples.

 We specify the brightness by dividing val by 4, because analogRead() returns a number up to 1023, and analogWrite() accepts a maximum of 255.

Try Other Analogue Sensors

The light-dependent resistor is a very useful sensor, but Arduino cannot directly read resistance. The circuit of Figure 5-6 takes the resistance of the LDR and converts it to a voltage that Arduino can read.

This same circuit works for any resistive sensor, and there are many different types of resistive sensors, such as sensors that measure force, stretching, bending, or heat. For example, you could connect a *thermistor* (heat-dependent resistor) instead of the LDR and have an LED that changes brightness according to the temperature.

If you do work with a thermistor, be aware that there isn't a direct connection between the value you read and the actual temperature measured. If you need an exact reading, you should read the numbers that come out of the analogue pin while measuring with a real thermometer. You could put these numbers side by side in a table and work out a way to calibrate the analogue results to real-world temperatures. Alternately, you could use a digital temperature sensor such as the Analog Devices TMP36.

Up to now, we have used an LED as the output device. It would be difficult to measure temperature, for instance, by trying to judge how bright an LED is. Wouldn't it be nice if we could actually get the values that Arduino is reading from the sensor? We could make the LED blink the values in Morse code, but there is a much easier way for Arduino to send information to us humans, using that same USB cable that you've been using to upload your sketches into the Arduino.

Serial Communication

You learned at the beginning of this book that Arduino has a USB connection used by the IDE to upload code into the microcontroller. The good news is that after a sketch is uploaded and is running, the sketch can use this same connection to send messages to or receive them from your computer. The way we do this from a sketch is to use the *serial object*.

An *object* is a collection of related capabilities bundled together for convenience, and the serial object allows us to communicate over the USB connection. You can think of the serial object as a cell phone, through which you have access to the cell-phone network. Like a cell phone, the serial object contains lots of complicated stuff that we don't need to worry about. We just need to learn how to use the serial object.

In this example you'll take the last circuit we built with the photoresistor, but instead of controlling the brightness of an LED, you'll send the values that are read from analogRead() back to the computer. Type the code in Example 5-5 into a new sketch. You can also download it from the example code link on the book's catalog page (*http://bit.ly/start_arduino_3e*).

Example 5-5. Send the computer the values read from analogue input 0

```
const int SENSOR = 0;  // select the input pin for the
                       // sensor resistor

int val = 0; // variable to store the value coming
             // from the sensor

void setup() {

  Serial.begin(9600); // open the serial port to send
                      // data back to the computer at
                      // 9600 bits per second
}

void loop() {

  val = analogRead(SENSOR); // read the value from
                            // the sensor

  Serial.println(val); // print the value to
                       // the serial port

  delay(100); // wait 100ms between
              // each send
}
```

After you've uploaded the code to your Arduino, you might think that nothing interesting happens. Actually, your Arduino is working fine: it is busy reading the light sensor and sending the information to your computer. The problem is that nothing on your computer is showing you the information that is coming from your Arduino.

What you need is the serial monitor, and it's built in to the Arduino IDE.

The Serial Monitor button is near the top-right corner of the Arduino IDE. It looks a bit like a magnifying glass, as if you were spying on the communication from the Arduino to your computer.

Click the Serial Monitor button to open the monitor, and you'll see the numbers rolling past in the bottom part of the window. Cover up the photoresistor to make it darker, and see how the numbers change. Notice that the numbers never go below zero, and never go above 1023, as this is the range of numbers that analogRead() can produce.

This serial communication channel between Arduino and your computer opens up a whole new world of possibilities. There are many programming languages that let you write programs for your computer that can talk to the serial port, and through the serial port, those programs can talk to your Arduino.

A particularly good complement to Arduino is the Processing language (*http://www.processing.org*)), because the languages and IDEs are so similar. You'll learn more about Processing in Chapter 7 in "Coding" on page 94, and your Arduino IDE includes some examples, such as File→Examples→04.Communication→Dimmer and File→Examples→04.Communication→Graph. You can also find many examples on the Internet.

Driving Bigger Loads (Motors, Lamps, and the Like)

Each of the pins on an Arduino board can only be used to power devices that use a very small amount of current, such as an LED. If you try to drive something big like a motor or an incandescent lamp, the pin might stop working, and could permanently damage the microcontroller that is the heart of your Arduino.

 To be safe, the current going through an Arduino I/O pin should be limited to 20 milliamps.

Don't worry, though. There are a number of simple techniques that allow you to control devices that use much more current. The trick is a bit like using a lever and fulcrum to lift a very heavy load. By putting a long stick under a big stone and a fulcrum in the right place, you can pull down on the long end of the stick, and the short end, under the stone, has much more force. You pull with a small force, and the mechanics of the lever apply a larger force to the stone.

In electronics, one way to do this is with a *MOSFET*. A MOSFET is an electronic switch that can be controlled by a small current, but in turn can control a much larger current. A MOSFET has three pins. You can think of a MOSFET as a switch between two of its pins (the *drain* and *source*), which is controlled by a third pin (the *gate*). It is a little like a light switch, where the gate is represented by the part you move to turn the light on and off. A light switch is mechanical, so it is controlled by a finger, but a MOSFET is electronic, so it is controlled by a pin from your Arduino.

 MOSFET means "metal-oxide-semiconductor field-effect transistor". It's a special type of transistor that operates on the field-effect principle. This means that electricity will flow though a piece of semiconductor material (between the *drain* and *source* pins) when a voltage is applied to the *gate* pin. As the gate is insulated from the rest through a layer of metal oxide, no current flows from Arduino into the MOSFET, making it very simple to interface. MOSFETs are ideal for switching on and off large loads at high frequencies.

In Figure 5-7, you can see how you would use a MOSFET like the IRF520 to turn on and off a motor attached to a fan. In this circuit, the motor actually takes its power from the VIN connector on the Arduino board, which is intended for a voltage between 7 and 12 volts. This is another benefit of the MOSFET: it allows us to control devices that need a different voltage from the 5 V used by Arduino.

The black component with the white band around it is a *diode*, and in this circuit it's being used to protect the MOSFET.

Conveniently, MOSFETs can be turned on and off very quickly, so you can still use PWM to control a lamp or a motor via a MOSFET. In Figure 5-7, the MOSFET is connected to pin 9, so you can use `analogWrite()` to control the speed of the motor through PWM.

To build the circuit, you will need an IRF520 MOSFET (*http:// bit.ly/ArduinoStoreIRF520*) and an 1N4007 diode (*http://bit.ly/ ArduinoStore1N4007*). If the motor randomly turns on during upload, place a 10 K ohm resistor between pin 9 and GND.

In Chapter 8 you'll learn about a *relay*, which is another way to control devices that use more current.

Complex Sensors

We define complex sensors as those that provide their information in a way that can't be read with `digitalRead()` or `analog-Read()` alone. These sensors usually have a whole circuit inside them, possibly with their own microcontroller. Some examples of complex sensors are digital temperature sensors, ultrasonic rangers, infrared rangers, and accelerometers. One reason for this complexity might be to provide more information or more accuracy; for example, some sensors have unique addresses, so you can connect many sensors to the same wires and yet ask each one individually to report its temperature.

Fortunately, Arduino provides a variety of mechanisms for reading these complex sensors. You'll see some of them in Chapter 8: in "Testing the Real Time Clock (RTC)" on page 110 to read a Real Time Clock, and in "Testing the Temperature and Humidity Sensor" on page 132, to read a Temperature and Humidity Sensor.

You can find more examples on our website in the Tutorials section (*http://bit.ly/11UYi7O*).

Tom Igoe's *Making Things Talk* (O'Reilly) has extensive coverage of complex sensors.

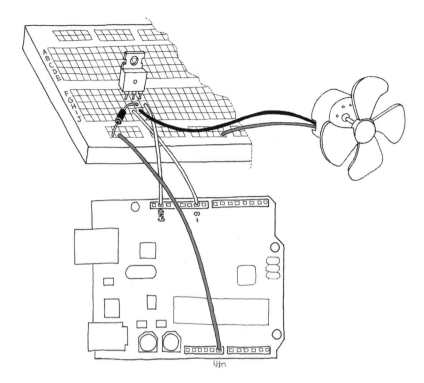

Figure 5-7. *A motor circuit for Arduino*

6/The Arduino Leonardo

Up to now, I've pretended there is only one type of Arduino board, the Arduino Uno. In fact, there are a number of different Arduino boards, which you can see by visiting the "Products" section of the Arduino website (*http://arduino.cc/en/Main/Prod ucts*).

In this chapter I want to discuss the Leonardo, because it's rather unique.

How Is This Arduino Different from All Other Arduinos?

As you might remember from Chapter 3, the heart of the Arduino Uno is a microcontroller called the ATmega328. What I didn't tell you then was that there is actually a second microcontroller on the Uno, an *ATmega16U2*, which is responsible for handling the USB interface. You may have noticed this next to the USB connector. The reason I haven't mentioned this chip before is that it's not easily accessible to beginners, and its job is to stay out of your way and handle the USB connection.

The reason the Arduino Uno needs this second microcontroller is that the ATmega328 can't process USB connections, while the ATmega16U2 doesn't have enough of the features to make a useful Arduino.

Fortunately, some microcontroller designers worked hard, and combined the Arduino features of the ATmega328 with the USB features of the ATmega16U2 to give us the ATmega32U4.

The Arduino Leonardo uses the ATmega32U4 to handle the USB work as well as all the Arduino work you're used to. This means that the code necessary to do the USB work is, in a way, part of your sketch, and this means that you can manipulate the USB behaviour from your sketch, using the special Mouse and Keyboard libraries. These libraries allow an Arduino Leonardo to appear as a mouse and/or keyboard to a connected computer. It turns out this allows you to do some pretty cool things.

 There are now other Arduino boards using the ATmega32U4, such as the Arduino Yun, the Arduino Micro, and the Arduino Esplora, as well as the Arduino robot. The features described here are applicable to all Arduinos based on the ATmega32U4.

Other Differences Between the Arduino Leonardo and the Arduino Uno

Before I show you how to do the cool things, there are a few other differences between the Arduino Uno and the Arduino Leonard.

The main differences that you should be aware of are these:

- In "Controlling Light with PWM" on page 56, you learned how to use analogWrite() to control the brightness of an LED. It turns out that not every Arduino pin can be used for this. On an Uno, analogWrite() works only with pins 3, 5, 6, 9, 10, and 11, while the Arduino Leonardo can also do analogWrite() on pin 13. (This means you can control the brightness of the built-in LED that is connected to pin 13!)

- When you connect an Arduino Uno to your computer, the USB serial connection is set up and stays there as long as the Arduino is connected to your computer, even if you reset your Arduino. Because the Arduino Leonardo creates the USB port *in the sketch*, resetting the Arduino Leonardo causes the USB serial connection to be broken and reestablished. Depending on what you're doing, this may have implications. If you're on Windows, it will probably make your computer beep a couple of times.

- The Arduino Uno has only 6 analogue inputs, A0–A5, which are all grouped together. The Leonardo has 12 analogue inputs, labeled A0 through A11. Inputs A0–A5 appear in the same locations as on the Uno, while inputs A6–A11 are on digital pins 4, 6, 8, 9, 10, and 12, respectively. These additional analogue input pins are labeled on the reverse side of the Leonardo.
- The Arduino Leonardo uses a Micro-B USB cable.
- The first time you plug a Leonardo into a Mac, the Keyboard Setup Assistant will launch. There's nothing to configure with the Leonardo, so close this dialogue by clicking the red button in the top left of the window.

Does it seem weird that some digital pins can also be used as analogue input pins? It turns out that all analogue pins can be used as digital pins, even on the Arduino Uno. You would refer to them by their analogue "name", for example:

```
pinMode(A4, OUTPUT);
```

or:

```
button = digitalRead(A3);
```

It's not unusual for microcontroller pins to have more than one use, although you can use the pin for only one of those abilities at a time.

Now that you know about the differences, let's make the Arduino Leonardo pretend to be a keyboard.

Don't forget to tell the Arduino IDE which board you're using! From the Tools menu in the IDE, select Board and then select the Leonardo.

Leonardo Keyboard Message Example

When the button is pressed in this example, a text string is sent to the computer as keyboard input. The string reports the number of times the button has been pressed.

Start by building the circuit. You'll need a simple pushbutton, just like the one you made in Figure 4-5, but use Arduino pin 4 instead of 7.

Next, open a new sketch in the Arduino IDE. You can name your sketch KeyboardMessage. Type the Example 6-1 code into Arduino. You can also download it from the example code link on the book's catalog page (*http://bit.ly/start_arduino_3e*) or open File→Examples→09.USB→Keyboard→KeyboardMessage in the IDE.

Example 6-1. Pretend to be a keyboard, and type a text string when a button is pressed

```
/*
Keyboard Button test

For the Arduino Leonardo and Micro.

Sends a text string when a button is pressed.

The circuit:
* pushbutton attached from pin 2 to +5V
* 10-kilohm resistor attached from pin 4 to ground

created 24 Oct 2011
modified 27 Mar 2012
by Tom Igoe

This example code is in the public domain.

http://www.arduino.cc/en/Tutorial/KeyboardButton
*/

const int buttonPin = 4;              // input pin for pushbutton
int previousButtonState = HIGH;       // for checking the state of a
pushButton
```

```
int counter = 0;                    // button push counter

void setup() {
  // make the pushButton pin an input:
  pinMode(buttonPin, INPUT);
  // initialize control over the keyboard:
  Keyboard.begin();
}

void loop() {
  // read the pushbutton:
  int buttonState = digitalRead(buttonPin);
  // if the button state has changed,
  if ((buttonState != previousButtonState)
    // and it's currently pressed:
  && (buttonState == HIGH)) {
    // increment the button counter
    counter++;
    // type out a message
    Keyboard.print("You pressed the button ");
    Keyboard.print(counter);
    Keyboard.println(" times.");
  }
  // save the current button state for comparison next time:
  previousButtonState = buttonState;
}
```

Upload the sketch, and then open a new document in any text editor or word processor. When you press the button, you should see something like this:

```
You pressed the button 1 times.
```

You might think that's not so exciting. Isn't this the same as sending a string to the serial monitor by using `Serial.println()`, as you did in Example 5-5? It's actually quite different.

When you send a string to the serial monitor by using `Serial.println()`, Arduino sends a special code for each letter (called the *ASCII code*, see "Variables" on page 216). On the computer side, the serial monitor is responsible for reading the ASCII code from the serial port and then telling your computer to display the correct letter. This works because the serial monitor knows what to do with serial ports.

In contrast, Example 6-1 is pretending to be a keyboard, and so the serial monitor is not needed. `Serial.println()` would not be able to type into a text editor or word processor, because they don't know how to read from a serial port.

Let's say that another way because this distinction is so important:

Using `Serial.println()` to send a message from Arduino to a program on your computer will work only with programs that will read a serial port, while Example 6-1 works with any program that expects input from a USB keyboard.

How Does This Work?

There's nothing special about the pushbutton part of the example, so I'll point out only the new concepts.

In `setup()`, the example initialises the keyboard object. Just like the serial object you learned about in "Serial Communication" on page 70, the keyboard object has a collection of capabilities bundled together.

In `loop()` the example waits for the button to be pressed, and then types the message by using `Keyboard.print()` and `Keyboard.println()`, which pretend that you typed the message out on a USB keyboard.

This means that your Arduino can send any key, not just characters like letters and numbers. In fact, in addition to keyboard keys, it can also send mouse clicks and mouse movements, which you'll see in the next example.

When you use the `Keyboard.print()` command, the Arduino Leonardo takes over your computer's keyboard. If the Keyboard (or Mouse) library is constantly running, it will be difficult to program your Leonardo.

To ensure that you don't lose control of your computer while running a sketch with these functions, always set up a reliable control system before you call `Keyboard.print()` or `Mouse.move()`. This sketch includes a pushbutton to toggle the keyboard, so that it runs only after the button is pressed.

Leonardo Button Mouse Control Example

When a button is pressed in this example, a mouse cursor movement is sent to the computer as if you had moved your mouse. Four buttons are used to send mouse movements of up, down, left, and right, and a fifth button is used to send a left mouse-button click.

Start by building the circuit. You'll need five pushbuttons, as shown in Figure 6-1.

As usual, open a new sketch in the Arduino IDE. Name your sketch ButtonMouseControl, and type the Example 6-2 code into Arduino. You can also download it from the example code link on the book's catalog page (*http://bit.ly/start_arduino_3e*), or open File→Examples→09.USB→Mouse→ButtonMouseControl in the IDE.

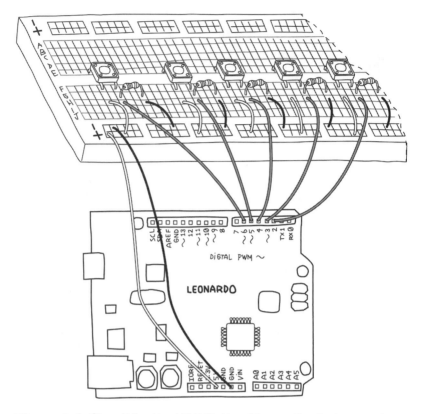

Figure 6-1. *Circuit for the USB Button Mouse Control example*

Example 6-2. Pretend to be a mouse, and send mouse movement events or a left mouse-click when the appropriate button is pressed

```
/*
  ButtonMouseControl

  Controls the mouse from five pushbuttons on an Arduino Leonardo
  or Micro.

  Hardware:
  * 5 pushbuttons attached to D2, D3, D4, D5, D6


  The mouse movement is always relative. This sketch reads
```

four pushbuttons, and uses them to set the movement of the mouse.

WARNING: When you use the Mouse.move() command, the Arduino takes
over your mouse! Make sure you have control before you use the mouse
commands.

created 15 Mar 2012
modified 27 Mar 2012
by Tom Igoe

this code is in the public domain

*/

```
// set pin numbers for the five buttons:
const int upButton = 2;
const int downButton = 3;
const int leftButton = 4;
const int rightButton = 5;
const int mouseButton = 6;

int range = 5;              // output range of X or Y movement;
affects
movement speed
int responseDelay = 10;     // response delay of the mouse, in ms

void setup() {
  // initialize the buttons' inputs:
  pinMode(upButton, INPUT);
  pinMode(downButton, INPUT);
  pinMode(leftButton, INPUT);
  pinMode(rightButton, INPUT);
  pinMode(mouseButton, INPUT);
  // initialize mouse control:
  Mouse.begin();
}

void loop() {
  // read the buttons:
  int upState = digitalRead(upButton);
  int downState = digitalRead(downButton);
  int rightState = digitalRead(rightButton);
  int leftState = digitalRead(leftButton);
  int clickState = digitalRead(mouseButton);
```

```
// calculate the movement distance based on the button states:
int  xDistance = (leftState - rightState)*range;
int  yDistance = (upState - downState)*range;

// if X or Y is non-zero, move:
if ((xDistance != 0) || (yDistance != 0)) {
  Mouse.move(xDistance, yDistance, 0);
}

// if the mouse button is pressed:
if (clickState == HIGH) {
  // if the mouse is not pressed, press it:
  if (!Mouse.isPressed(MOUSE_LEFT)) {
    Mouse.press(MOUSE_LEFT);
  }
}
// else the mouse button is not pressed:
else {
  // if the mouse is pressed, release it:
  if (Mouse.isPressed(MOUSE_LEFT)) {
    Mouse.release(MOUSE_LEFT);
  }
}

// a delay so the mouse doesn't move too fast:
delay(responseDelay);
}
```

Upload the sketch, and then press the movement buttons (the ones attached to pins 2, 3, 4, or 5) one at a time. You should see the cursor on your computer move around. Move the cursor over a partially hidden window and press the button that corresponds to clicking the left mouse-button. The hidden window should be brought to the front.

How Does This Work?

The important functions here are Mouse.move(), Mouse.press(), and Mouse.release().

Apart from the fact that there are five buttons, Example 6-2 is more complicated than Example 6-1 because the amount of mouse movement has to be calculated. The sketch uses a step size of 5, and then either adds or subtracts this amount to the variable xDistance depending on whether the left or right button is pressed, and either adds or subtracts this amount to the variable yDistance depending on whether the up or down button is pressed. Once the movement amount is calculated, the cursor is moved using Mouse.move(xDistance, yDistance, 0).

If the fifth button, the one that corresponds to clicking the left mouse-button, is pressed, the Leonardo sends a mouse-press message using Mouse.press(MOUSE_LEFT).

As you can see, this means you can control any program on your computer with the Leonardo. You can create your own physical "shortcut" button that automatically composes and sends an email message to a friend that says, "Sorry I'll be late for lunch; see you in half an hour". You can build a pretend airplane cockpit with various switches, joysticks, and knobs and use it to play your favourite video game. You can attach a big red button to your Leonardo that tells your computer to use the Arduino IDE to reprogram the Leonardo. You can even attach a light sensor to your Leonardo and have it log you out of your computer when it gets dark, forcing you to go home. (See File→Examples→09.USB→Keyboard→KeyboardLogout.)

The last two examples point out that you should use this new power very carefully. Anything that you can do using your mouse and keyboard can now be done from your Leonardo, including deleting files, rebooting your computer, and changing passwords. Make sure you understand what you are doing before becoming too adventurous.

You can learn more about the Leonardo USB Keyboard and Mouse libraries and see examples on the Arduino website (*http://arduino.cc/en/Reference/MouseKeyboard*).

More Leonardo Differences

There are other differences that as a beginner you are less likely to encounter, but I'll mention them here so that you're aware of them. Most of the differences have to do with ports being on different pins.

Just as your computer might have different types of ports (USB, video, 1394, parallel), your Arduino has different ports. You've already been using the serial port. Here are the others:

- Another type of port is something called I2C, which we use in Chapter 8. Both the Uno and the Leonardo have an I2C port, but it's on different pins. On the Arduino Uno, the I2C port is on analogue pins A4 and A5, while on the Leonardo the I2C port is on digital pins D2 and D3.

  To avoid confusion, the Arduino Uno R3, Arduino Leonardo, and all newer Arduinos duplicate the I2C pins above the AREF pins. These new pins are labeled SCL and SDA and will always be in the same place, above the AREF pin, regardless of what type of microntroller the board uses.

- All Arduino boards have a special way to program brand-new microcontrollers (for example, in case you have to replace the microcontroller). This is done using another special port called ICSP.

  As you might expect, it's on different pins on the Uno and the Leonardo. However, unlike the I2C port, the ICSP port already had its own connector: it's the group of six pins arranged in two rows of three pins.

  However, some older libraries and shields, designed *before* the Leonardo existed, made use of the fact that the ICSP port was also on known digital pins, and those libraries and shields will not work properly with the Leonardo because the ICSP port is not present on those pins. (Shields are boards that plug into the pins of an Arduino and provide additional functionality.)

- In "Serial Communication" on page 70 you learned how to use the serial port to send sensor information from the Arduino to your computer. You won't be surprised to know that in addition to going to the USB connector, the serial port on the Arduino Uno is also present on digital pins 0 and 1.

  While the Uno has only the one serial port, the Leonardo has two serial ports, and the one connected to digital pins 0 and 1 is called *Serial1*, while on the Uno it's called *Serial*. The serial port that goes to the USB connector is always named *Serial* on all Arduino boards.

There are a few differences that don't have to do with ports:

- When you refer to an analogue input pin using the names A0–A5, those names get translated into numbers, and all the analogue and digital functions work the same whether you use the name or the number. For example, on the Uno, saying:

```
pinMode(A0,OUTPUT);
```

is exactly the same as saying:

```
pinMode(14,OUTPUT);
```

On the Arduino Uno, the analogue pins are numbered sequentially following D13, so 14 is the same as A0, 15 the same as A1, etc.

On the Leonardo the pins are numbered differently, so any library or example that uses the numbers instead of the names will not work properly. From now on, libraries and examples should always use the analogue name of the pin, even when being used in digital modes, e.g.:

```
digitalWrite(A4, HIGH);
```

- The Arduino Uno *always* resets when you open the serial monitor. Some libraries and examples take advantage of this, but because the Leonardo does not reset when the serial monitor is opened, those libraries may not work properly.

- An advanced Arduino capability called *external interrupts* behaves differently on the Leonardo compared to the Uno. The Uno has only two external interrupts and the Leonardo has five, and while the Leonardo has interrupts on the same two pins as on the Uno, they are numbered differently.

You can learn more about the Arduino Leonardo on the Arduino website (*http://arduino.cc/en/Guide/ArduinoLeonardoMicro*).

7/Talking to the Cloud

In the preceding chapters, you learned the basics of Arduino and the fundamental building blocks available to you. Let's go over what makes up the "Arduino Alphabet":

Digital output
> We used it to control an LED but, with the proper circuit, it can be used to control motors, make sounds, and a lot more.

Analogue output
> This gives us the ability to control the brightness of the LED, not just turn it on or off. We can even control the speed of a motor with it.

Digital input
> This allows us to read the state of sensors that just say yes or no, like pushbuttons or tilt switches.

Analogue input
> We can read signals from sensors that have more information than just on or off, like a potentiometer that can tell where it's been turned to, or a light sensor that can tell how much light is on it.

Serial communication
> This allows us to communicate with a computer and exchange data or simply monitor what's going on with the sketch that's running on the Arduino.

In this chapter, we're going to see how to put together a working application using what you have learned in the previous chapters. This chapter should show you how every single example can be used as a building block for a complex project.

> Here is where the wannabe designer in me comes out. We're going to make the twenty-

first-century version of a classic lamp by my
favourite Italian designer, Joe Colombo. The
object we're going to build is inspired by a lamp
called Aton from 1964.

—Massimo

The lamp, as you can see in Figure 7-1, is a simple sphere sitting
on a base with a large hole to keep the sphere from rolling off
your desk. This design allows you to orient the lamp in different
directions.

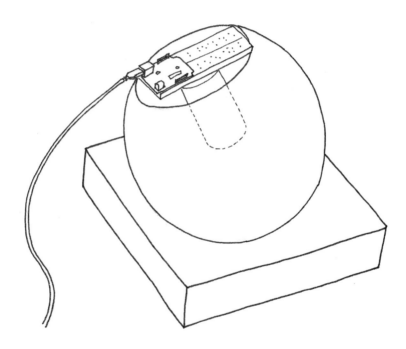

Figure 7-1. *The finished lamp*

In terms of functionality, we want to build a device that would
connect to the Internet, fetch the current list of articles on the
Make blog (*http://blog.makezine.com*), and count how many
times the words "peace," "love," and "Arduino" are mentioned.
With these values, we're going to generate a colour and display

it on the lamp. The lamp itself has a button we can use to turn it on and off, and a light sensor for automatic activation.

Planning

Let's look at what we want to achieve and what bits and pieces we need. First of all, we need Arduino to be able to connect to the Internet. As the Arduino board has only a USB port, we can't plug it directly into an Internet connection, so we need to figure out how to bridge the two. Usually what people do is run an application on a computer that will connect to the Internet, process the data, and send Arduino some simple bit of distilled information.

Arduino is a simple computer with a small memory; it can't process large files easily, and when we connect to an RSS feed, we'll get a very verbose XML file that would require a lot more RAM. On the other hand, your laptop or desktop computer has much more RAM and is much better suited for this kind of work, so we'll implement a proxy to simplify the XML using the Processing language to run on your computer.

Processing

Processing is where Arduino came from. We love this language and use it to teach programming to beginners as well as to build beautiful code. Processing and Arduino are the perfect combination. Another advantage is that Processing is open source and runs on all the major platforms (Mac, Linux, and Windows). It can also generate standalone applications that run on those platforms. What's more, the Processing community is lively and helpful, and you can find thousands of premade example programs.

Get Processing from *https://processing.org/download*.

The proxy does the following work for us: it downloads the RSS feed from *http://makezine.com* and extracts all the words from the resulting XML file. Then, going through all of them, it counts the number of times "peace," "love," and "Arduino" appear in the text. With these three numbers, we'll calculate a colour value and send it to Arduino. The Arduino code, in turn, will send to the computer the amount of light measured by the sensor, which the Processing code will then display on the computer screen.

On the hardware side, we'll combine the pushbutton example, the light sensor example, the PWM LED control (multiplied by 3!), and serial communication. See if you can identify each of these circuits when you build it in "Assembling the Circuit" on page 101. This is how typical projects are made.

As Arduino is a simple device, we'll need to codify the colour in a simple way. We'll use the standard way that colours are represented in HTML: # followed by six hexadecimal digits.

Hexadecimal numbers are handy, because each 8-bit number is stored in exactly two characters; with decimal numbers this varies from one to three characters. Predictability also makes the code simpler: we wait until we see a #, and then we read the six characters that follow into a *buffer* (a variable used as a temporary holding area for data). Finally, we turn each group of two characters into a byte that represents the brightness of one of the three LEDs.

Coding

There are two sketches that you'll be running: the Processing sketch and the Arduino sketch. Example 7-1 is the code for the Processing sketch. You can also download it from the example code link on the book's catalog page (*http://bit.ly/start_ardu ino_3e*).

Example 7-1. Arduino Networked Lamp

Parts of the code are inspired by a blog post by Tod E. Kurt (*http://todbot.com*).

```
import processing.serial.*;
import java.net.*;
import java.io.*;
import java.util.*;

String feed = "http://makezine.com/feed/";
```

```
int interval = 5 * 60 * 1000;  // retrieve feed every five
minutes;
int lastTime;                  // the last time we fetched the
content

int love   = 0;
int peace  = 0;
int arduino = 0;

int light = 0;  // light level measured by the lamp

Serial port;
color c;
String cs;

String buffer = ""; // Accumulates characters coming from Arduino

PFont font;

void setup() {
  size(640, 480);
  frameRate(10);    // we don't need fast updates

  font = createFont("Helvetica", 24);
  fill(255);
  textFont(font, 32);

  // IMPORTANT NOTE:
  // The first serial port retrieved by Serial.list()
  // should be your Arduino. If not, uncomment the next
  // line by deleting the // before it, and re-run the
  // sketch to see a list of serial ports. Then, change
  // the 0 in between [ and ] to the number of the port
  // that your Arduino is connected to.
  //println(Serial.list());
  String arduinoPort = Serial.list()[0];

  port = new Serial(this, arduinoPort, 9600); // connect to
Arduino

  lastTime = millis();
  fetchData();
}

void draw() {
```

```
background( c );
int n = (lastTime + interval - millis())/1000;

// Build a colour based on the 3 values
c = color(peace, love, arduino);
cs = "#" + hex(c, 6); // Prepare a string to be sent to Arduino

text("Arduino Networked Lamp", 10, 40);
text("Reading feed:", 10, 100);
text(feed, 10, 140);

text("Next update in "+ n + " seconds", 10, 450);
text("peace", 10, 200);
text(" " + peace, 130, 200);
rect(200, 172, peace, 28);

text("love ", 10, 240);
text(" " + love, 130, 240);
rect(200, 212, love, 28);

text("arduino ", 10, 280);
text(" " + arduino, 130, 280);
rect(200, 252, arduino, 28);

// write the colour string to the screen
text("sending", 10, 340);
text(cs, 200, 340);

text("light level", 10, 380);
rect(200, 352, light/10.23, 28); // this turns 1023 into 100

if (n <= 0) {
  fetchData();
  lastTime = millis();
}

port.write(cs); // send data to Arduino

if (port.available() > 0) { // check if there is data waiting
  int inByte = port.read(); // read one byte
  if (inByte != 10) { // if byte is not newline
    buffer = buffer + char(inByte); // just add it to the buffer
  } else {

    // newline reached, let's process the data
    if (buffer.length() > 1) { // make sure there is enough data
```

```
          // chop off the last character, it's a carriage return
          // (a carriage return is the character at the end of a
          // line of text)
          buffer = buffer.substring(0, buffer.length() -1);

          // turn the buffer from string into an integer number
          light = int(buffer);

          // clean the buffer for the next read cycle
          buffer = "";

          // We're likely falling behind in taking readings
          // from Arduino. So let's clear the backlog of
          // incoming sensor readings so the next reading is
          // up-to-date.
          port.clear();
        }
      }
    }
}

void fetchData() {
  // we use these strings to parse the feed
  String data;
  String chunk;

  // zero the counters
  love    = 0;
  peace   = 0;
  arduino = 0;
  try {
    URL url = new URL(feed);  // An object to represent the URL
    // prepare a connection
    URLConnection conn = url.openConnection();
    conn.connect(); // now connect to the Website

    // this is a bit of virtual plumbing as we connect
    // the data coming from the connection to a buffered
    // reader that reads the data one line at a time.
    BufferedReader in = new
      BufferedReader(new
InputStreamReader(conn.getInputStream()));

    // read each line from the feed
    while ( (data = in.readLine()) != null) {
```

```
StringTokenizer st =
  new StringTokenizer(data, "\"<>,.()[] ");// break it down
while (st.hasMoreTokens ()) {
  // each chunk of data is made lowercase
  chunk= st.nextToken().toLowerCase() ;

  if (chunk.indexOf("love") >= 0 ) // found "love"?
    love++;     // increment love by 1
  if (chunk.indexOf("peace") >= 0)   // found "peace"?
    peace++;    // increment peace by 1
  if (chunk.indexOf("arduino") >= 0) // found "arduino"?
    arduino++; // increment arduino by 1
  }
}

// Set 64 to be the maximum number of references we care
about.
  if (peace > 64)   peace = 64;
  if (love > 64)    love = 64;
  if (arduino > 64) arduino = 64;
  peace = peace * 4;     // multiply by 4 so that the max is
255,
  love = love * 4;       // which comes in handy when building a
  arduino = arduino * 4; // colour that is made of 4 bytes
(ARGB)
  }
catch (Exception ex) { // If there was an error, stop the sketch
  ex.printStackTrace();
  System.out.println("ERROR: "+ex.getMessage());
  }
}
```

There is one thing you need to do before the Processing sketch
will run correctly: you need to confirm that the sketch is using
the correct serial port for talking to Arduino. You'll need to wait
until you've assembled the Arduino circuit and uploaded the
Arduino sketch before you can confirm this. On some systems,
this Processing sketch will run fine. However, if you don't see
anything happening on the Arduino and you don't see any information from the light sensor appearing onscreen, find the
comment labeled IMPORTANT NOTE in the Processing sketch and
follow the instructions there.

If you're on a Mac, there's a good chance your Arduino will be on the last serial port in the list. If so, you can replace the 0 in `Serial.list()[0]` with `Serial.list().length -1`. This subtracts one from the length of the list of all serial ports; array indexes count from zero, but `length` tells you the size of the list (counting from one), so you need to subtract one to get the actual index.

Example 7-2 is the Arduino sketch. You can also download it from the example code link on the book's catalog page (*http://bit.ly/start_arduino_3e*).

Example 7-2. Arduino Networked Lamp (Arduino sketch)

```
const int SENSOR = 0;
const int R_LED = 9;
const int G_LED = 10;
const int B_LED = 11;
const int BUTTON = 12;

int val = 0; // variable to store the value coming from the sensor

int btn = LOW;
int old_btn = LOW;
int state = 0;
char buffer[7] ;
int pointer = 0;
byte inByte = 0;

byte r = 0;
byte g = 0;
byte b = 0;

void setup() {
  Serial.begin(9600);  // open the serial port
  pinMode(BUTTON, INPUT);
}

void loop() {
```

```
val = analogRead(SENSOR); // read the value from the sensor
Serial.println(val);      // print the value to
                          // the serial port

if (Serial.available() > 0) {

  // read the incoming byte:
  inByte = Serial.read();

  // If the marker's found, next 6 characters are the colour
  if (inByte == '#') {

    while (pointer < 6) { // accumulate 6 chars
      buffer[pointer] = Serial.read(); // store in the buffer
      pointer++; // move the pointer forward by 1
    }

    // now we have the 3 numbers stored as hex numbers
    // we need to decode them into 3 bytes r, g and b
    r = hex2dec(buffer[1]) + hex2dec(buffer[0]) * 16;
    g = hex2dec(buffer[3]) + hex2dec(buffer[2]) * 16;
    b = hex2dec(buffer[5]) + hex2dec(buffer[4]) * 16;

    pointer = 0; // reset the pointer so we can reuse the buffer

  }
}

btn = digitalRead(BUTTON); // read input value and store it

// Check if there was a transition
if ((btn == HIGH) && (old_btn == LOW)){
  state = 1 - state;
}

old_btn = btn; // val is now old, let's store it

if (state == 1) { // if the lamp is on

  analogWrite(R_LED, r);  // turn the leds on
  analogWrite(G_LED, g);  // at the colour
  analogWrite(B_LED, b);  // sent by the computer
} else {

  analogWrite(R_LED, 0);  // otherwise turn off
  analogWrite(G_LED, 0);
```

```
  analogWrite(B_LED, 0);
  }

  delay(100);                    // wait 100ms between each send
}

int hex2dec(byte c) { // converts one HEX character into a number
    if (c >= '0' && c <= '9') {
      return c - '0';
    } else if (c >= 'A' && c <= 'F') {
      return c - 'A' + 10;
    }
}
```

Assembling the Circuit

Figure 7-2 shows how to assemble the circuit. Just as you did in "Controlling Light with PWM" on page 56 in Chapter 5, use a 220-ohm resistor (red-red-brown) with each LED, and just as you did in "Analogue Input" on page 66, use a 10 K ohm resistor with the photoresistor.

Remember from "Controlling Light with PWM" on page 56 that LEDs are polarized: in this circuit, the anode (long lead, positive) should go to the right, and the cathode (short lead, negative) to the left. Figure 7-2 also shows the flattened side of the LED, which indicates the cathode.

Build the circuit as shown, using one red, one green, and one blue LED. Next, load the sketches into Arduino and Processing. Upload the Arduino sketch to the Arduino, and then run the Processing sketch and try it out (you will need to press the button to get the lamp to come on). If you run into any problems, check Chapter 9.

Instead of using three separate LEDs, you can use a single RGB LED, which has four leads coming off it. You'll hook it up in much the same way as the LEDs shown in Figure 7-2, with one change: instead of three separate connections to the ground pin on Arduino, you'll have a single lead (called the *common cathode*) going to ground.

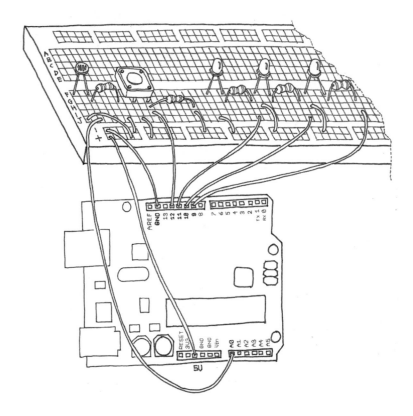

Figure 7-2. *The Arduino Networked Lamp circuit*

The Arduino Store sells a four-lead RGB LED for a few dollars. Also, unlike discrete single-colour LEDs, the longest lead on this kind of RGB LED is the one that goes to ground. The three shorter leads will need to connect to Arduino pins 9, 10, and 11 (with a 220-ohm resistor between the leads and the pins, just as with the separate red, green, and blue LEDs).

The Maker Shed Getting Started with Arduino Kit (*http://bit.ly/ get-started-arduino-v3*) includes an RGB LED as well.

The Arduino sketch is designed to work with a *common cathode* RGB LED (one where the long lead goes to ground). If you're getting the wrong output, you might have a *common anode* RGB LED. If that's the case, change the code where you set the LED intensity as shown (you are basically inverting the values; where you used 0, you'd now use 255):

```
if (state == 1) { // if the lamp is on
    analogWrite(R_LED, 255 - r);  // turn the leds
on
    analogWrite(G_LED, 255 - g);  // at the colour
    analogWrite(B_LED, 255 - b);  // sent by the
computer
} else {
    analogWrite(R_LED, 255);  // otherwise turn off
    analogWrite(G_LED, 255);
    analogWrite(B_LED, 255);
}
```

Now let's complete the construction by placing the breadboard into a glass sphere. The simplest and cheapest way to do this is to buy an IKEA FADO table lamp. It's now selling for about US $19.99/€14.99/£11.99 (ahh, the luxury of being European).

Here's How to Assemble It

Unpack the lamp and remove the cable that goes into the lamp from the bottom. You will no longer be plugging this into the wall.

You can use a rubber band to strap the Arduino to the breadboard, and then hot-glue the breadboard onto the back of the lamp, as shown in Figure 7-1. Leave some room so that you can insert the LED and glue it in place.

Solder longer wires to the RGB LED and glue it where the lightbulb used to be. Connect the wires coming from the LED to the breadboard (where it was connected before you removed it). You can save a bit of time by noting that you will need only one connection to ground, whether you're using the RGB LED or three separate LEDs.

Now find a nice piece of wood with a hole that can be used as a stand for the sphere, or just cut the top of the cardboard box that came with the lamp at approximately 5 cm (or 2″) and make a hole with a diameter that cradles the lamp. Reinforce the inside of the cardboard box by using hot glue all along the inside edges, which will make the base more stable.

Place the sphere on the stand and bring the USB cable out of the top and connect it to the computer.

Fire off your Processing code, press the On/Off button, and watch the lamp come to life. Invite your friends over and amaze them!

As an exercise, try to add code that will turn on the lamp when the room gets dark. Other possible enhancements are as follows:

- Add tilt sensors to turn the lamp on or off by rotating it in different directions.
- Add a PIR sensor to detect when somebody is around, and turn it off when nobody is there to watch.
- Create different modes so that you can get manual control of the colour or make it fade through many colours.

Think of different things, experiment, and have fun!

8/Automatic Garden-Irrigation System

In Chapter 7 you combined what you had learned about Arduino into a project, the Arduino Networked Lamp. Part of the fun was to combine some of the simple exercises into a practical project. You also learned about the Processing language and how to use it to create a proxy on your computer to do things that would be difficult or impossible with your Arduino.

In this chapter you will again combine simple examples with some new ideas to make a practical project. Along the way you'll learn more about electronics, communication, and programming, and we'll give some attention to construction techniques.

The goal of this project is to automatically turn the water on and off at the right time each day, except if it's raining.

 If you don't have a garden, you can still have some fun with this project. If you just have a small house-plant you want to water, try building this with only one valve. If you need to dispense a tasty beverage of your choice at 5 p.m. each day, consider using a food-grade pump instead of the water valve. For example, Adafruit sells a Peristaltic Liquid Pump with Silicone Tubing (*http://bit.ly/15oTtpH*).

As a professor, I teach many students to build things. Along the way it occurred to me that students sometimes think that I instantly know exactly how to build a project. In fact, designing a project is an extremely iterative process.

—Michael

To create a project, start with an idea, and rough out little pieces of it; as you go, this sometimes requires making changes to the initial idea. We often have to take a detour to learn how a new electronic part works, or to figure out a programming concept we've not encountered before, or remind ourselves how to use a feature of Arduino we've not used in a long time or is new to us. Sometimes we have to turn to our textbooks, the Internet, or ask someone for advice. We review many examples, tutorials, and projects that contain bits related to what we are doing. We take bits and pieces from different places and combine them, perhaps very roughly at first, like Frankenstein's monster, to see how things will work together.

As the project progresses from concept to rough design to testing parts of the hardware and software, we keep having to go back and make changes in something we did earlier, so that everything will work together properly. We don't know a single engineer who starts with a blank piece of paper, designs a whole project from start to finish, which then works exactly as planned, without ever having to go back and change anything.

All of the preceding is true for hardware as well as software.

The point here is that even if you are a beginner, you are *ready* to design projects. Start with what you know, and slowly add features, one new idea or part at a time. Don't be afraid to explore intriguing ideas that have no immediate use.

> Whenever I hear of an electronic part or programming concept or trick that seems interesting, I try it out, even if I don't have a use for it right away. This knowledge then becomes another tool in my toolkit. If you get stuck or don't know something, remember that even professional engineers have to learn new things all the time.
>
> —Michael

Thanks to the wide and generous Arduino community, you have many resources via the Internet, and unless you are a hermit on a mountain top you can probably find a local Arduino meetup, club, makerspace, hackerspace, or even individual who can help.

For some hints on how to make the best use of online resources, see "How to Get Help Online" on page 203.

So, in addition to teaching you more about electronics, programming, and construction, I'm going to show you a bit about the design process. You'll see that some of the simplistic circuits or sketches will get modified again and again until we arrive at the final project. Even so, I've skipped over some iteration steps in order to keep this chapter from becoming an entirely new book. Iterations take time!

Planning

As in Chapter 7, start by thinking about what you want to achieve and what bits and pieces you'll need.

This project will use common gardening electric water valves, available in home improvement stores. While at the store, you will also need one power supply, or transformer, suitable for these water valves. In Chapter 5 you learned how to use a MOS-FET to control a motor. This might work for the water valves, except that some water valves might use *alternating current* (AC), and MOSFETs can control only *direct current* (DC). In order to control AC, you need a *relay*, which can control both AC and DC.

 In "Driving Bigger Loads (Motors, Lamps, and the Like)" on page 72 in Chapter 5, you learned that a MOSFET is a type of transistor, in which the *gate* pin can control whether electricity flows between the *drain* and *source* pins. In this sense, a MOSFET is a switch. A relay is also a switch. Inside the relay is a tiny mechanical switch controlled by an electromagnet: by turning on and off the electromagnet, you control whether electricity flows through the mechanical switch.

In order to know when to turn the water on and off, we'll need a clock of some kind. You could try to do this in your program using the Arduino built-in timer, but it would be complicated, and worse, it's not terribly accurate. As it turns out, a device that does this exists, is quite inexpensive, and is easy to use with Arduino. The RTC (Real Time Clock) is similar to the device in your computer that keeps track of the date and time even if you leave it turned off for a long time.

We'll also need a sensor to tell us if it's raining. We'll use a Temperature and Humidity Sensor, as they are inexpensive and easy to use. We don't need to know the temperature, but you get that extra feature "for free" and it might be useful.

Finally, we'll need a way to set the on-times and off-times, i.e., the *user interface*. To keep this project from getting out of hand, I'll use the serial monitor for the user interface. As you become more fluent in Arduino, you could replace this with an LCD display and pushbuttons.

Before you start programming, you need to think about how the hardware will be connected. I like to use a rough block diagram to help me see all the parts I need and to think through how they should be connected. Eventually, you'll need to know exactly how to connect things, but in the block diagram (Figure 8-1) we just use one line to symbolize some kind of connection.

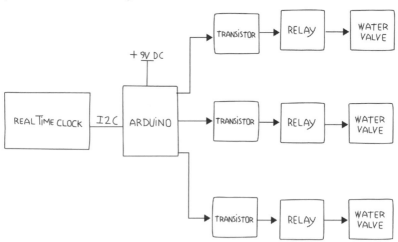

Figure 8-1. *Block diagram showing Real Time Clock, Arduino, MOSFETs, relays, valves, and power supplies*

In this diagram we assumed three separate water valves, but you can see how this can be generalized to whatever your needs are.

As this is a more advanced project, I'll introduce construction techniques. This project must work reliably for many months, perhaps even years, so you've got a different goal from a simple example that is only meant to show you how something works. The solderless breadboard you used earlier is great for proto-typing or experimenting, but for reliability we'll build this project by soldering components to a *Proto Shield*. We'll also consider how to distribute power and make all the connections to the various external parts like the water valves. We'll even look at how this project could be protected in some kind of enclosure.

 Shields are boards that plug into the pins of an Arduino and provide additional functionality. The Arduino Proto Shield is a particular shield that is designed to let you build your own circuit on it.

Another feature that is useful as your projects get more complex is an indication of what is going on. This is helpful for debugging, and is especially useful when parts of your system are far away, such as the water valves. We'll add LEDs to indicate that the water valves are activated. Don't forget the resistors for the LEDs.

Now that we have a few more details, I like to make a tentative shopping list. With complex systems, I expect that I'll have to make changes: for example, as I work on the sketch, I might realise that I need another part. (The final, complete shopping list, with links, is at "Irrigation Project Shopping List" on page 191.)

Don't worry if you don't know what all these parts are. We'll cover them in detail as we go:

- One Real Time Clock (RTC)
- One Temperature and Humidity Sensor
- One Proto Shield

- Three electric water valves
- One transformer or power supply for the water valves
- Three relays to control the water valves
- Three sockets for relays
- Three LEDs as valve activation indicators
- Three resistors for LEDs
- One power supply for Arduino (so it will work even when a computer isn't attached)

Now that you have a tentative list, let's look at each item and work out the details. Let's start with the RTC.

Testing the Real Time Clock (RTC)

When I plan to use a device that is new to me, I like to first verify that I understand how it works, before designing the whole system. Because the RTC is new to you, let's take a look at how it works.

The main part of an RTC is the chip itself. The most common one is the DS1307, which requires a crystal for correct timekeeping and a battery to keep it running when the rest of the system is switched off. Rather than build this ourselves, we'll use one of the many RTC modules available, saving us time for very little cost.

DS1307 RTC modules are available from many different sources. Fortunately, they all function and interface in very similar ways. I ended up with the TinyRTC, available from oddWires (*http://bit.ly/11t1Huu*). It's also available on the Maker Shed as part #MKAD19 (*http://bit.ly/1vJSThM*).

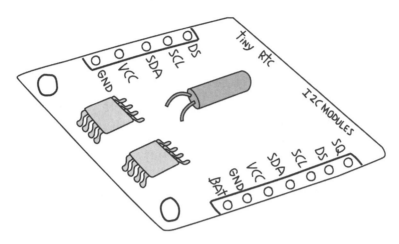

Figure 8-2. *TinyRTC Real Time Clock module*

Note that the male headers are not always included; you may have to order those separately and solder them in. You can get male headers from many sources; for example, Adafruit part #392 (*http://bit.ly/1ycvkwR*) includes plenty of male headers for this and future projects.

 If you're new to soldering, there is a link to a great soldering tutorial in "Soldering Your Project on the Proto Shield" on page 170.

This device uses an interface called I2C, sometimes also called the Two Wire Interface (TWI) or simply Wire. Arduino provides a built-in library for this called Wire, and Adafruit provides a library for the DS1307 (*https://github.com/adafruit/RTClib*).

To use this library, click the Download ZIP button, and then unzip this file in the folder named *libraries* in your Arduino sketch folder.

 You can find the sketch folder by opening Arduino's preferences (File→Preferences, or Arduino→Preferences on Mac) and looking at the field labeled Sketchbook Location.

When you unzip it, the library will be sitting in a folder named *RTClib-master*, which you have to rename to *RTClib*.

If you have the Arduino IDE open, you will need to close and reopen it in order to recognise the new library.

You can check that you've installed the library correctly by checking the example that comes with this library. You don't need to build a circuit for this, and technically you don't even need to have your Arduino handy. In the Arduino IDE, open the File menu and select Examples→RTClib→ds1307 to open an example program. Instead of clicking the Upload button, click the Verify button (see Figure 4-2). If you get the message "Done compiling," then you have installed the library correctly.

 After installing a new library, it's wise to verify that the library is installed properly, and that any libraries it depends on are also installed properly, before you try writing your own program.

Most Arduino libraries include examples, and because they were probably written by the same people who wrote the library, they are most likely correct.

The TinyRTC module comes with two sets of pins: one consists of five positions and the other of seven positions. Most of the pins are duplicated, and other pins provide additional features. To test the RTC, you only need to worry about four pins: the two pins that form the I2C interface, power, and ground; the RTC must be provided with 5 V on its pin labeled VCC; and its ground must be connected to the Arduino ground.

 On the hardware side, I2C is supported by two specific Arduino pins (SDA and SCL), as described in the Arduino Wire Library reference (*http://arduino.cc/en/reference/wire*).

For a quick test, you can make use of a common trick: for devices that use very little power, such as the RTC, you can use the digital outputs to provide power by setting one pin to HIGH and another to LOW, if the pins line up properly. An I/O pin that is set to output HIGH is essentially the same as 5 V, and an I/O pin that is set to output LOW is essentially the same as the ground.

On an Uno, SCL is A5 and SDA is A4. In addition to these two interface pins, we need to connect VCC and GND to any other I/O pins to provide 5V and GND. This can be accomplished by aligning the TinyRTC as shown in Figure 8-3 on the Arduino Uno analogue input pins. You could also use a breadboard and jumper wires to make the connection.

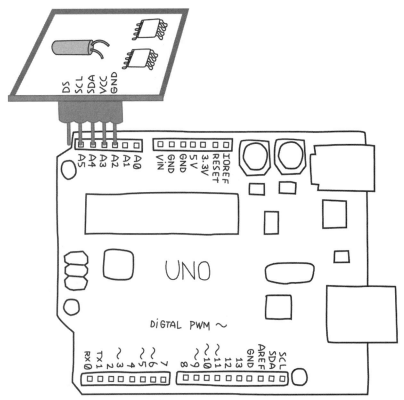

Figure 8-3. *TinyRTC plugged directly into Arduino Uno analogue input pins. The offset is intentional and enables SCL to plug into A5, and SDA to A4.*

Note that on recent Arduinos with different micro-controllers, for example the Arduino Leonardo, the I2C signals (SDA and SCL) could be on different pins. For this reason, all new Arduinos now comply with a new standard pin layout, which adds the I2C pins after AREF. This is in addition to and duplicates whatever other pins happen to be the I2C pins.

Analogue input A4 and A5 will deal with I2C communication, while A2 and A3 will provide power and ground. A3 needs to provide 5 V to the RTC pin labeled VCC, so we will set it HIGH, while A2 will be set LOW to provide ground.

Now we are ready to test! In the Arduino IDE, open the File menu and select Examples→RTClib→ds1307 to open the example program. Before you compile and upload, remember that you need to set up pins A2 and A3 to deliver power to the TinyRTC. Add the following four lines to **setup()**, right at the very top:

```
void setup() {
  pinMode(A3, OUTPUT);
  pinMode(A2, OUTPUT);
  digitalWrite(A3, HIGH);
  digitalWrite(A2, LOW);
```

 If you are using a board that's wired differently, you'll need to use a breadboard and jumper wires to connect it, so you should not add those extra lines of code. Just be sure to wire the board as instructed by the supplier.

You can also download the example with this modification from the example code link on the book's catalog page (*http://bit.ly/start_arduino_3e*).

While here, note that the example opens the serial port at 57,600 baud.

Now you can upload, and after the upload has completed, open the serial monitor. Check the baud rate selection box in the lower-right corner of the serial monitor, and select "57600 baud". You should see output resembling this:

```
2013/10/20 15:6:22
  since midnight 1/1/1970 = 1382281582s = 15998d
  now + 7d + 30s: 2013/10/27 15:6:52

2013/10/20 15:6:25
  since midnight 1/1/1970 = 1382281585s = 15998d
  now + 7d + 30s: 2013/10/27 15:6:55
```

Note that the date and time may be wrong, but you should see the seconds count increasing. If you get an error, double-check that the RTC is in the right position, i.e., that SCL is connected to A5 etc., and double-check that you set pins A2 and A3 properly in setup().

To set the correct time in the RTC, look in the setup() function. Towards the end you will see this line:

```
rtc.adjust(DateTime(__DATE__, __TIME__));
```

This line takes the date and time at which the sketch was compiled (__DATE__ and __TIME__, respectively) and uses that to set the RTC. Of course, it might be off by a second or two, but that's close enough for our purposes.

Copy that line so that it's outside the if() condition, e.g., just below the rtc.begin():

```
rtc.begin();
rtc.adjust(DateTime(__DATE__, __TIME__));
```

Compile and upload the sketch, and now the serial monitor should display the correct date and time:

```
014/5/28 16:12:35
  since midnight 1/1/1970 = 1401293555s = 16218d
  now + 7d + 30s: 2014/6/4 16:13:5
```

Of course, if your computer has the wrong date and time, this will be reflected here.

After you set the time on the RTC you should comment out the rtc.adjust (and upload the code), otherwise you will keep resetting the time to when that sketch was compiled. This RTC will now keep time for years.

For more information on the library and examples, read Arduino Library (*http://bit.ly/1vJTRuI*) and Understanding the Code (*http://bit.ly/11t22xq*) on Adafruit's tutorial (*http://bit.ly/1xSkpbl*) for their DS1307 Breakout Board kit.

Note that while the Adafruit board is different, the code is the same.

Now that you feel comfortable with the RTC, let's move on to the relays.

Testing the Relays

What kind of relays do we need? It depends on how much current the water valves need. Most garden valves seem to use 300 milliamps. This is a small amount of current, and so a small relay is enough. Relays that can be operated at different voltages are available; we'll use one that uses 5 V so we don't need another power source. Figure 8-4 shows a popular small 5 V relay.

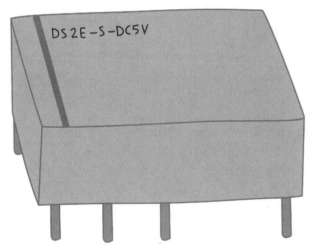

Figure 8-4. *5 V relay*

You can buy it at Digi-Key (*http://bit.ly/1xAspKl*), as well as from many other vendors.

Almost every electronic device has what's called a *data sheet*, where all the detailed technical information on the device is documented. It can be a little overwhelming for the beginner, as there is so much information, and usually you need only a very tiny part of it. As you get more experienced, you'll learn what's important and how to find it quickly. If you search for the data sheet for the relay we've chosen, you'll see that it can handle up to 2 amps and 30 volts of direct current (DC), or 1 amp and 125 volts of alternating current (AC), which is more than enough for us. This relay also has the advantage of being compatible with our solderless breadboard as well as the Proto Shield you will use later.

Whenever you want to control something with an Arduino output, you have to remember that an Arduino pin should power only devices that use up to 20 milliamps (see "Driving Bigger Loads (Motors, Lamps, and the Like)" on page 72). If you search for the current used by this relay in the data sheet, you won't find it. However, you will find the resistance. Now you have to do some math, because knowing the resistance of the relay (125 ohms) and voltage that Arduino puts out on the I/O pins (5 V), you can calculate the current by using Ohm's law, which you learned about near the end of "What Is Electricity?" on page 37. Dividing the voltage (5) by the resistance (125), we get the current: 40 milliamps.

Since that's over our limit, we'll need MOSFETs. For a change, we'll use a different MOSFET from the one we used in "Driving Bigger Loads (Motors, Lamps, and the Like)" on page 72. We'll use the 2N7000, and you can find its data sheet posted on the Fairchild Semiconductor website (*http://bit.ly/1vck9Cd*).

Just as in "Driving Bigger Loads (Motors, Lamps, and the Like)" on page 72, the gate will be controlled by the Arduino I/O pin, and the drain and source will be the switch that will control the relay. You'll have to to add three 2N7000 MOSFETs to the shopping list, one for each relay.

To avoid the MOSFET gates from floating, add three 10 K ohm resistors to your shopping list, one for each relay.

When you turn on or reset an Arduino, all the digital pins start off as inputs, until your program starts running and your `pinMode()` changes any pins to outputs. This is important because in that brief period of time, before the `pinMode()` function changes your pin to an output, the gate will be neither HIGH nor LOW: it will be left *floating*, which means that the MOSFET could easily turn on, causing the water to come on briefly. While this might not be the end of the world in most projects, it's a good habit to account for this. As hinted at in "Driving Bigger Loads (Motors, Lamps, and the Like)" on page 72 in Chapter 5, a 10 K ohm between the I/O pin and ground prevents this. 10 K ohms is a low enough resistance to make sure the gate doesn't "float", but it's also a large enough resistance that it won't get in our way when we want to turn on the water.

A resistor used this way is called a *pull-down* resistor, because it "pulls" the gate down to ground. Sometimes a connection needs to be "pulled" to 5V; in this case, it's called a *pull-up* resistor.

Whenever we control a relay or motor, we should add a diode to protect the MOSFETs from the *flyback voltage* generated by the collapsing magnetic field when the relay is turned off. Although our MOSFET has a built-in diode, it is a relatively small one, so for reliability, it is wise to add an additional, external diode. So, another addition to your shopping list, this time for three 1N4148 (or similar) diodes. While we're at it, we should add the relay part number, and because we know the type of relay, we can also indicate the correct socket for the relay. Here are the additions that take the shopping list to what we'll call revision 0.1:

- Add three MOSFETs to control the relays, 2N7000
- Add three resistors, 10 K ohm
- Add three diodes, 1N4148 or equivalent

- Add three relays, DS2E-S-DC5V

Sounds like this circuit is getting complex, doesn't it? It's hard to visualise how all the components are supposed to be connected to each other.

Fortunately, a clever system exists for capturing this information. It's called a *schematic diagram*.

Electronic Schematic Diagrams

Most electronic circuits are completely defined by two things: 1) which components are used and 2) how they are to be connected. By capturing only this information as clearly as possible, a schematic diagram is the clearest way to visualise and communicate an electronic circuit.

A schematic diagram intentionally does not convey the size, shape, or colour of components, or how they are physically placed next to each other, because this information is not relevant to the definition of the circuit; rather, these are construction details for a particular implementation of that circuit.

Each component is represented by a schematic symbol, which unambiguously identifies the component, but says nothing about its size, colour, etc.

In some cases, schematic symbols look very much like the components they represent, while in other cases they are quite different. In particular, you will see that the schematic symbol for an Arduino does not look at all like an Arduino. From the point of view of a schematic diagram, the only thing that matters about an Arduino is that it has certain pins (power, inputs, outputs, etc.). Thus, it is drawn as a very plain box with just the pins indicated.

Electronic schematic symbols and diagrams are designed to convey their functions as quickly and clearly as possible, and some conventions have become common. Two of the most important conventions are as follows:

- The lowest voltage is shown at the bottom of the schematic diagram, and the highest voltage is at the top. Usually this means that the GND connection is at the bottom and that 5V (or a higher voltage if that's used) is at the top.

- The information flows from left to right. Thus sensors and other input devices are shown on the left, and outputs such as motors, LEDs, relays, and water valves are on the right. If the information travels from Arduino to the MOSFET to a relay to a water valve, they will be shown on a schematic diagram in that order from left to right, even though when you build your circuit, you will have different priorities and your layout may be quite different.

The schematic symbol for an Arduino reflects these conventions too: VIN, 5V, and 3V3 are at the top, GND is at the bottom, and the various controls (RESET, AREF, etc.) are on the left because they are inputs to the Arduino. Although Arduino has three GND pins, only one is shown on the schematic symbol since they are electrically identical. Because the digital and analogue pins can be either inputs or outputs, their placement is somewhat arbitrary.

You can learn more about schematic diagrams in Appendix D.

Back to the project at hand: Figure 8-5 shows the schematic diagram for the circuit we've been discussing. Remember, the goal is to verify that we can control a relay by using a MOSFET (although the finished system will have three water valves, for the purpose of verifying that the plan works, we need to check only one).

Note the pin numbers next to the schematic symbol of the relay. These are important because they tell you which pin is connected to what inside. It is critical to understand this in order to wire up your circuit properly. Note that the pin spacing is not equal: pins 1 and 4 are farther apart than pins 4 and 8. Note that there is a black stripe on the top between pins 8 and 9. Finally, note that the pin numbers are as viewed from the bottom.

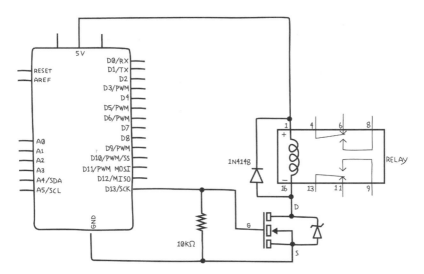

Figure 8-5. *Circuit schematic diagram for testing Arduino control of a relay*

For reference, Figure 8-6 is a picture of the same circuit on a solderless breadboard in the style we've used up to now. You might think of this as a pictorial circuit diagram, as opposed to the schematic in Figure 8-5.

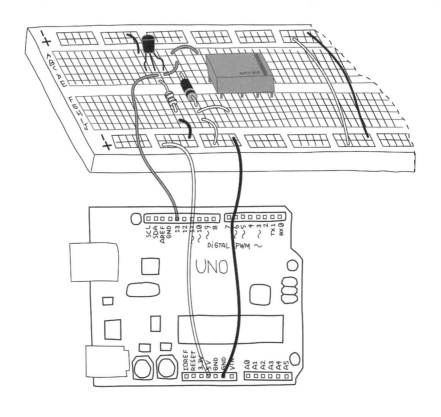

2N7000

Figure 8-6. *Pictorial circuit diagram for testing Arduino control of a relay*

Just like the relay, the MOSFET pins must be properly identified and wired up correctly. Note that the MOSFET has a curved side and a flat side, as shown in Figure 8-6. This is necessary to indicate the proper order of the pins. Be aware that this order is particular to the 2N7000, but is not necessarily universal. Other MOSFETs might have a different order of the pins. You always have to check the data sheet to find the order of pins for your particular MOSFET.

Pay attention to the diodes, MOSFET, and the relay: the diodes are polarized, the MOSFET must have the flat side facing the right way, and the relay needs to have the stripe at the proper end for everything to work.

If you look at Figure 8-5 and Appendix D, you'll notice that some components, like resistors, light sensors, and (some) capacitors, have circuit symbols that are *symmetrical*, in that you can flip them upside down and they look exactly the same, while other components, like LEDs, diodes, and MOSFETs, are not symmetrical. Resistors, light sensors, and (some) capacitors are *unpolarized*, meaning they work the same regardless of which way the current flows through them, while components, like LEDs and diodes, are *polarized*, meaning they work differently depending on which way the current flows through them. Similarly, MOSFET pins have very specific functions, and can't be used interchangeably. While not uniformly true, generally components that have symmetrical schematic symbols are unpolarized, and components with asymmetrical schematic symbols are polarized.

Once you've built the circuit, the next step is to write the sketch. For testing, I prefer to use one of Arduino's built-in examples, if possible, since I know the sketch is correct. Because the relay will make a faint clicking sound when activated, if you run the

Blink example, you should hear the relay clicking once a second, and you won't have to write a line of code.

Before uploading the sketch, verify that the sketch is controlling the pin that is attached to the MOSFET. If you have followed the schematic, you will be connected to pin 13, which is in fact the pin that is controlled in the Blink example sketch. Additionally, the LED will blink at the same time the relay should click.

 Before you upload a sketch, it's a good habit to verify that the pins used in the sketch are indeed the pins that you have wired up. You might have a perfectly correct sketch, and a perfectly built circuit, but if your sketch uses different pins than your circuit, things won't work, and you might waste lots of time trying to find the problem.

If you don't hear the relay clicking, check the troubleshooting hints in Chapter 9. Remember, the clicking is very faint; you'll need to put your ear very close to the relay and be in a very quiet room.

Now we can add the water valve. The water valve will connect to the relay and its own power supply. The water valve and power supply probably come with stranded wire, which is almost impossible to use with a solderless breadboard. You'll find it handy to attach a short piece of solid-core wire to the stranded wire when you're working with a solderless breadboard, as shown in Figure 8-7.

Figure 8-7. *Short piece of solid-core wire soldered to stranded wire for solderless breadboard work*

You'll need to insulate the joint with some electrical tape or heat shrink tubing to prevent it from touching a wire that it shouldn't.

Whenever you have an exposed piece of metal, such as the wires you just joined together, or the long, bare leads of a photoresistor, or even something that is not part of the circuit, like a screw, you need to make sure that it can't touch other parts of a circuit, making a connection that you don't want. This is called a *short circuit*, and it can make your project fail to work properly. To prevent short circuits, always insulate any exposed wires or secure things in such a way that they can't move and touch something that they shouldn't.

You are probably already familiar with electrician's or insulating tape, which is a common, inexpensive, and easy method. A more professional technique is to use heat shrink tubing. The tubing is cut to size and slipped over the exposed wires or connections, and then heated up with a heat gun, which causes the tubing to shrink tightly around the connection.

At this point you might realise that you'll need a way to make this connection when you build the final system. There are many ways to do this, but good-quality screw terminals are a great choice, as shown in Figure 8-8.

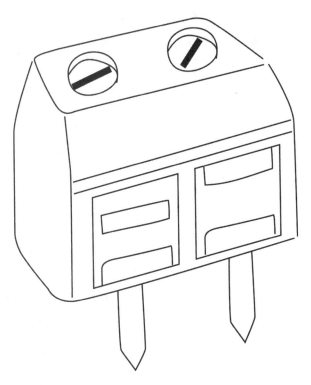

Figure 8-8. *Good-quality screw terminal with two positions*

So you have another addition to your shopping list. Here are the additions to take us to revision 0.2 of the shopping list:

- Four two-position screw terminals (one pair for each water valve, plus one pair for the water valve power supply) e.g., Jameco (*http://www.jameco.com*) part no. 1299761

Figure 8-9 shows the circuit schematic with the water valve and power supply added.

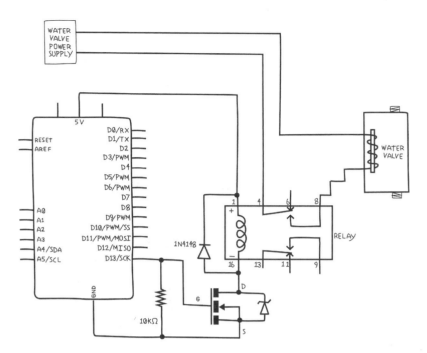

Figure 8-9. *Circuit schematic for testing Arduino control of one water valve*

And Figure 8-10 shows the pictorial circuit diagram of exactly the same circuit.

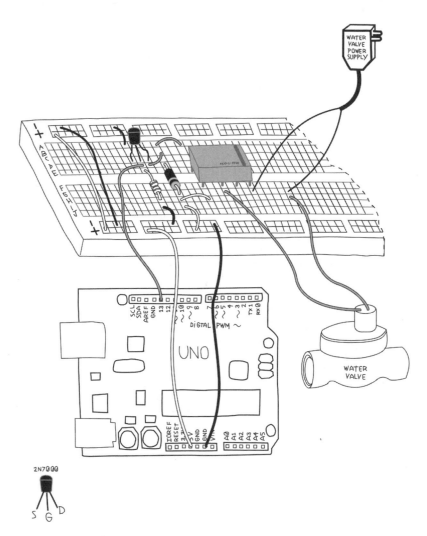

Figure 8-10. *Pictorial circuit diagram for testing Arduino control of one water valve*

The order of the water valve and the water valve power supply are changed in these two views. I did this to avoid having wires cross in the diagrams for clarity. When connecting a component and a power supply via a switch, the order doesn't matter, as long as the switch controls whether the circuit is closed or open.

You can use the same Blink sketch; you should still hear the relay clicking. You might not hear the water valve clicking, because some water valves work only when they have water pressure inside of them. I was fortunate; my water valves made a very loud click when they were turned on.

What about the LEDs? They can be installed in a number of different places: at the digital outputs, at the MOSFET output, or at the relay output. When possible, I like to put the LED on the farthest point, to verify as much as possible, so let's put it at the relay output. (If you want, you can put LEDs at all mentioned stages, making it very easy to spot exactly where the signal stops.)

What resistor should we use? This LED will be getting power from the water valve power supply. Most water valves seem to be either 12 or 24 volts. For safety, let's design for a 24 V system, and if yours is a 12 V system, you can either reduce the value or have a slightly dimmer LED. It should still be quite visible.

```
LED resistor = (voltage of supply - voltage of LED)/(desired
LED current)
```

 If you're uncertain about what value resistor to use when you're using it to limit current, it's always safer to choose a bigger-value resistor. The LED will be visible over quite a wide range of values, and, if it's too dim, you can always reduce the resistance.

Most LEDs use about 2 volts and are safe below 30 mA, so we have $R = (24-2)V/30mA = 733$ ohms. You can safely round this up to 1 K ohm; the result will be a little less current and a slightly dimmer LED.

But wait, in "Planning" on page 107, I told you that some water valves use AC, and then later in "Electronic Schematic Diagrams" on page 119, I told you that LEDs are polarized. *Polarized* means that the LEDs care which way the current flows, and *AC* means that the current changes direction all the time. Won't that damage the LEDs? As it turns out, LEDs can withstand a certain amount of voltage in the wrong direction, but if the voltage is too high, the LED might be damaged. Fortunately, other diodes can withstand much higher voltages safely, so let's use another three 1N4148 diodes to protect the LEDs, giving us an addition that takes the shopping list to revision 0.3:

- Change quantity of 1N4148 or similar diodes to six
- Specify the LED resistor value is 1 k ohm

Figure 8-11 shows the circuit schematic with the LED and diode added. We've indicated the polarity of the water valve power supply and of the water valve, but this is relevant only if you have a DC system. If you have an AC system (which seems to be more common), these have no polarity.

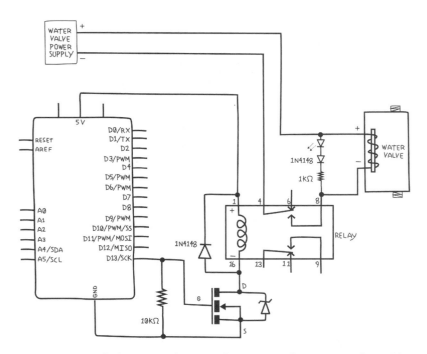

Figure 8-11. *Schematic diagram for testing the water valve with added LED*

And Figure 8-12 shows the pictorial circuit diagram. I've left off the resistor values, so make sure you compare with the schematic and get the right resistor in the right place. Also pay attention to the LED and diode polarity—the LED anode connects to the water valve and the diode anode connects to the LED cathode.

Before you plug in the water valve power supply, double check your wiring, especially the relay connections. You don't want the water valve power supply voltage to get into the Arduino, as it would almost certainly cause damage. Once again, run the Blink sketch. In addition to hearing the relay and possibly the valve clicking, you should see the LED light up.

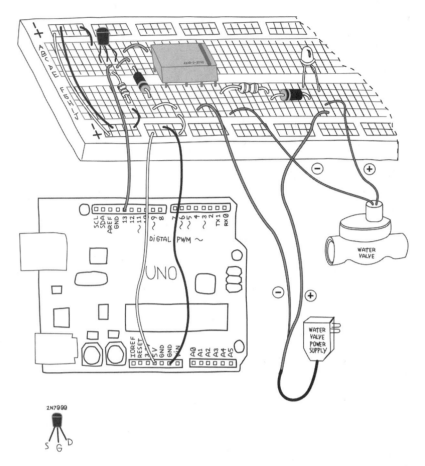

Figure 8-12. *Pictorial circuit diagram for testing the water valve with added LED*

Now that we have the relay and water valve figured out, let's test the Temperature and Humidity Sensor.

Testing the Temperature and Humidity Sensor

The DHT11 is a popular Temperature and Humidity Sensor. Like the RTC, it is inexpensive and easy to use with Arduino. According to the data sheet, the DHT11 is connected as shown in Figure 8-13. Note the pullup resistor on the data pin.

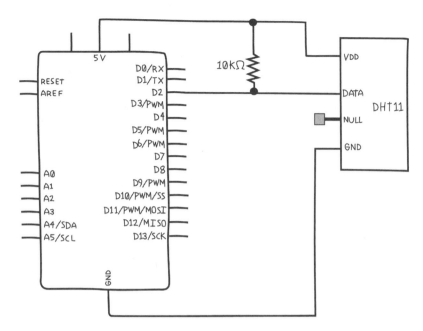

Figure 8-13. *Schematic diagram for testing the DHT11 Temperature and Humidity Sensor*

Because we're adding a component that needs one, let's add another 10 K ohm resistor to our shopping list. We're now at version 0.4:

- Add one resistor, 10 K ohm (for Temperature and Humidity Sensor)

Because of the pull-up resistor, we can't use the trick we did with the RTC (snapping it directly onto the Arduino), so we'll have to put this on a breadboard (Figure 8-14).

 The schematic diagram of a circuit is the same regardless of whether you build the circuit on a breadboard or in some other way.

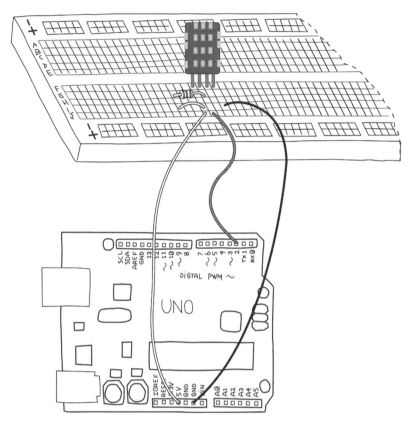

Figure 8-14. *Pictorial circuit diagram for testing the DHT11 Temperature and Humidity Sensor*

You can download the DHT11 library from GitHub (*http://bit.ly/ 1trMqke*).

As you did with the RTC library, click the Download ZIP button, and then unzip this file in the folder named *libraries* in your Arduino sketch folder. The library will be created in a folder named *DHT-sensor-library-master*, which you have to rename to *DHT*. Remember that if you have the Arduino IDE open, you will need to restart it to make it aware of the new library. Check that you've properly installed the library by opening the DHTtester example in the DHT category of examples, and then click the Verify button (see Figure 4-2). If you get the message "Done compiling", you have installed the library correctly.

Before you upload the sketch to your Arduino, note that the example supports three different models of DHT sensors: DHT11, DHT21, and DHT22. To correctly select the proper model, a constant named DHTTYPE is defined to be either DHT11, DHT21, or DHT22:

```
// Uncomment whatever type you're using!
//#define DHTTYPE DHT11   // DHT 11
#define DHTTYPE DHT22   // DHT 22  (AM2302)
//#define DHTTYPE DHT21   // DHT 21 (AM2301)
```

Notice that DHT11 and DHT21 are ignored, because those two lines are comments or, as programmers say, those lines are *commented out*. Because you are using a DHT11 sensor, you need to *comment out* the line with the DHT22, and "uncomment" the line with the DHT11:

```
// Uncomment whatever type you're using!
#define DHTTYPE DHT11   // DHT 11
//#define DHTTYPE DHT22   // DHT 22  (AM2302)
//#define DHTTYPE DHT21   // DHT 21 (AM2301)
```

The lines with the DHT22 and the DHT21 aren't doing anything, but they serve to remind us that the library will work with these three sensors, and that this is how you specify which one you are using.

You may have encountered another type of constant: the *constant variable*. Apart from the awkward name, it too is important and useful.

The differences between a *variable* (like an integer), a *constant variable*, and a *named constant value* are subtle and a bit complicated. In broad terms, a *constant variable* uses a tiny bit of your Arduino's memory and obeys *scope rules*. In contrast, a *named constant value* does not use any memory and always has global scope.

As a general rule, you should avoid using *named constant values* and use them only if a library requires them.

You can learn more about *named constant values* (*http://bit.ly/1uUiDpC*), the `const` keyword (*http://bit.ly/1vjSHDK*), and *variable scope* (*http://bit.ly/1zqEKmM*) on the Arduino website.

Once you have defined the correct type of DHT sensor, verify that the sketch uses the same pin that you have connected to the sensor, and upload the DHTtester example to your Arduino and open the serial monitor. You should see something like this:

```
DHTxx test!
Humidity: 47.00 % Temperature: 24.00 *C 75.20 *F Heat index:
77.70 *F
Humidity: 48.00 % Temperature: 24.00 *C 75.20 *F Heat index:
77.71 *F
```

You can check the humidity sensor by gently exhaling on it. The moisture in your breath should make the humidity rise. You can try to raise the temperature by placing your fingers around the sensor, but since you're touching the plastic case and not the sensor itself, you may not be able to raise the temperature by much.

Now that you can feel confident in your components, you can start designing the software.

Coding

Guess what? Writing code (*coding*) requires planning as well. You need to think a little about what you are trying to do before you start typing away. Similar to the way you tested the new electronics before doing the whole design, you'll test each piece of code before going on. The less code there is, the easier it is to find problems.

Setting the On and Off Times

We want to turn the water valves on and off at different times of the day. We'll need some way to record those values. Because we have three valves, we might use an array, with one entry for each valve. This will also make it easier if we later want to add more water valves. You might recall that we used an array named **buffer** (see Example 7-2 in Chapter 7) to store the characters as they were sent from the Processing sketch. Arrays are described briefly in "Variables" on page 216.

Here's one way to do that:

```
const int NUMBEROFVALVES = 3;
const int NUMBEROFTIMES = 2;

int onOffTimes [NUMBEROFVALVES][NUMBEROFTIMES];
```

 For simplicity, we'll assume you turn the water on and off at the same time every day of the week. As your programming skills increase, you can modify this to accommodate different schedules on different days of the week, and even multiple times in one day. When you start a project, it's good to start with the most simple system possible, and add features as you verify that things work properly.

Note that instead of using fixed numbers for the dimensions of the array, I first created two constant variables. This serves to remind me what these numbers mean, and make it easier to change later. Constant variable names are in uppercase letters to remind us that they are constants.

If you've never seen a two-dimensional array, don't be frightened. Think of it as a spreadsheet. The number in the first [] is the number of rows, and the number in the second [] is the number of columns. A row represents a valve, and we'll use the first column to store the time at which to turn the valve on, and the second column to store the time at which to turn the valve off.

Let's make constant variables for the column numbers as well. Remember that the index of elements within an array always starts at zero:

```
const int ONTIME = 0;
const int OFFTIME = 1;
```

Next we need a way to input this information; that is, some sort of *user interface*. Typically, a user interface is a menu, but we're going to make something extremely simple using the serial monitor.

Remember in Chapter 7 we needed a way to tell Arduino what colour to make the light? As I mentioned there, because Arduino is a simple device, we chose a simple way to codify the colour.

We're going to do a similar thing here: codify the times in as simple a way as possible.

We need to be able to set the ON time and OFF time for each valve. We could use a number indicating the desired valve followed by the letter *N* for "on" and *F* for "off," followed by the time. We could input the time in 24-hour notation, e.g., 0135 for 1:35 a.m. Thus, we would type

2N1345 2F1415

to turn valve 2 on at 1:45 p.m. and off at 2:15 p.m.

To make our life easier, let's insist that we always use uppercase letters for *N* and *F*.

In our code, we would need to *parse*, or separate, the string that we type into the correct groups.

 A group of consecutive characters is called a *string*.

If you look at the Arduino sketch (Example 7-2), you'll see we made use of `Serial.available()` and `Serial.read()`, which are functions of the serial object. It turns out the serial object has more functions, as described at Arduino (*http://bit.ly/1uCgj1v*).

We're going to make use of the `Serial.parseInt()` function, which automatically reads any incoming digits and converts them to a number. It stops when it sees a character that isn't a number. The letters (*N* or *F*) we'll read directly with the `Serial.read()` function.

For the purposes of testing, we'll just print out the entire array after each line is read, as shown in Example 8-1.

Example 8-1. Parsing the commands sent to the irrigation system

```
const int NUMBEROFVALVES = 3;
const int NUMBEROFTIMES = 2;

int onOffTimes [NUMBEROFVALVES][NUMBEROFTIMES];

const int ONTIME = 0;
const int OFFTIME = 1;

void setup(){
  Serial.begin(9600);
};

void loop() {
  // Read a string of the form "2N1345" and separate it
  // into the first digit, the letter, and the second number

  // read only if there is something to read
  while (Serial.available() > 0) {

    // The first integer should be the valve number
    int valveNumber = Serial.parseInt();

    // the next character should be either N or F
    // do it again:
    char onOff = Serial.read();

    // next should come the time
```

```arduino
  int desiredTime = Serial.parseInt();
  //Serial.print("time = ");
  //Serial.println(desiredTime);

  // finally expect a newline which is the end of
  // the sentence:
  if (Serial.read() == '\n') {
    if ( onOff == 'N') { // it's an ON time
      onOffTimes[valveNumber][ONTIME] = desiredTime;
    }
    else if ( onOff == 'F') { // it's an OFF time
      onOffTimes[valveNumber][OFFTIME] = desiredTime;
    }
    else { // something's wrong
      Serial.println ("You must use upper case N or F only");
    }
  } // end of sentence
  else {
    Serial.println("no Newline character found"); // Sanity
check
  }

  // now print the entire array so we can see if it works
  for (int valve = 0; valve < NUMBEROFVALVES; valve++) {
    Serial.print("valve # ");
    Serial.print(valve);
    Serial.print(" will turn ON at ");
    Serial.print(onOffTimes[valve][ONTIME]);
    Serial.print(" and will turn OFF at ");
    Serial.print(onOffTimes[valve][OFFTIME]);
    Serial.println();
  }
  } // end of Serial.available()
}
```

You can download this sketch from this book's catalog page (*http://bit.ly/start_arduino_3e*).

After loading the sketch into Arduino, open the serial monitor and check the baud rate and line-ending selection boxes in the lower-right corner of the serial monitor. Select Newline and 9600 baud. The line-ending type makes sure that every time you end a line by pressing the Enter key on your computer, your computer will send a newline character to your Arduino.

For instance, if you want to turn on valve #1 at 1:30 p.m., type **1N1330** and press Enter. You should see this:

```
valve # 0 will turn ON at 0 and will turn OFF at 0
valve # 1 will turn ON at 1330 and will turn OFF at 0
valve # 2 will turn ON at 0 and will turn OFF at 0
```

In the sketch, notice that I check to make sure that the character between the numbers is either *N* or *F*, and that after the second number is a newline character. This sort of "sanity checking" is handy to catch mistakes you might make while typing, which might confuse your program. These sort of checks are also useful for catching mistakes in your program. You might think of other sanity checks; for instance, you might check that the time is valid, i.e., it is less than 2359, and that the valve number is less than **NUMBEROFVALVES**.

A program that is designed to handle only the correct data is very delicate, as it is sensitive to any errors, whether caused by user input or a mistake elsewhere. By checking the data before operating on it, your program will be able to identify errors instead of trying to operate with faulty data, which could lead to unexpected or erroneous behavior. This makes your program robust, which is a highly desireable quality, especially since humans won't always do what they are supposed to do.

Before we go on, I want to show you a new trick. This chunk of code we just developed is lengthy, and we still have a lot to add. It's going to start getting confusing to read and manage the program.

Fortunately, we can make use of a very clever and common programming technique. In "Pass Me the Parmesan" on page 31 we explained what a function is, and that **setup()** and **loop()** are two functions that Arduino expects to exist. So far you've created your entire program within these two functions.

What I didn't emphasize is that you can create other functions the same way you create **setup()** and **loop()**.

Why is this important? Because it's a very convenient way to break a long and complicated program into small functions, each of which has a specific task. Furthermore, since you can name those functions anything you want, if you use names that describe what they do, reading the program becomes much simpler.

 Any time a chunk of code that does a specific task becomes large, it is a good candidate for becoming a function. How large is large enough? That's up to you. My rule of thumb is that as soon as a chunk of code takes up more than two screens, it is a candidate for becoming a function. I can keep two screenfuls in my head, but not more than that.

An important consideration is whether the chunk of code can easily be extracted. Does it depend on many variables that are visible only within another function? As you learn more about *variable scope*, you'll see this is important too.

For instance, the code we just developed reads a command from the serial monitor, parses it, and then stores the times into the array. If we made this a function called `expectValveSetting()`, our loop is simply:

```
void loop() {

    expectValveSettings();

}
```

That's much easier to read, and most important, easier to understand as we develop the rest of the program.

Of course we need to create this function, which we do like this:

```
void expectValveSettings() {
    // Read a string of the form "2N1345" and separate it
    // into the first digit, the letter, and the second number

    // read only if there is something to read
    while (Serial.available() > 0) {
```

```
// ... rest of the code not repeated here
}
```

The rest is exactly the same as Example 8-1 ; I left out the rest of the function because I didn't want to waste another two pages.

Now we can turn to the other things we need to do, and make them functions as well.

Checking Whether It's Time to Turn a Valve On or Off

Next, let's look at the data from the RTC and figure out how we'll use this to decide whether it's time to turn something on or off. If you go back to the RTC example *ds1307*, you'll see how the time is printed:

```
Serial.print(now.hour(), DEC);
```

Conveniently, this is already a number, so comparing with the hours and minutes we have stored will be easy.

In order to access the RTC, you'll need to add parts of the *ds1307* example to your program. At the top, before **setup()**, add this:

```
#include <Wire.h>
#include "RTClib.h"

RTC_DS1307 rtc;
```

This time we won't use the analogue input pins to provide 5V and GND. Why? Because analogue inputs are scarce; there are only six of them, and we already lost two for the I2C interface to the RTC. At the moment our project doesn't need any analogue inputs, but we might think of something later.

In **setup()** you'll need the following:

```
#ifdef AVR
    Wire.begin();
#else
    Wire1.begin(); // Shield I2C pins connect to alt I2C
bus on Arduino Due
```

```
#endif
    rtc.begin();
```

Now think about what you need to do: as long as the current time is greater than the time you want to turn a valve ON, and less than the time you want to turn the valve OFF, you should turn it on. At all other times you want it turned OFF.

From the RTC library you can get the current time like this:

```
dateTimeNow = rtc.now();
```

and then you can access parts of the time like this:

```
dateTimeNow.hour()
dateTimeNow.minute()
```

Can you see the problem? We've stored time as a four-digit number, where the first two digits are the hour and the last two digits are the minutes. We can't do a mathematical comparison without separating this number into hours and minutes, and then the comparison gets complicated.

It would be nice if we could have just one number. We can do this if, instead of storing the time as hours and minutes, we just converted the hours to minutes and stored the number of minutes since midnight. That way, we only have to deal with one number and the mathematical comparison is easy. (We have to remember never to turn on the water before midnight and off after midnight. This would be another opportunity for a sanity check to make our program more robust.)

The following code illustrates this:

```
int nowMinutesSinceMidnight = (dateTimeNow.hour() * 60) +
    dateTimeNow.minute();
```

and then the comparison looks like this:

```
if ( ( nowMinutesSinceMidnight >= onOffTimes[valve]
[ONTIME]) &&
    ( nowMinutesSinceMidnight < onOffTimes[valve]
[OFFTIME]) ) {
        digitalWrite(??, HIGH);
}
else {
        digitalWrite(??, LOW);
}
```

Wait a minute, what about those question marks? We need to know the pin number of each valve. Our `for()` loop just counts off the valves: 0, 1, and 2. We need a way to indicate what pin number corresponds to which valve. We can use an array:

```
int valvePinNumbers[NUMBEROFVALVES];
```

By using the same constant variable we created earlier, this array will always have exactly the same number of rows as the other array, even if you later change the number of valves.

In `setup()` we'll insert the correct pin numbers into the array:

```
valvePinNumbers[0] = 6; // valve 0 is on pin 6
valvePinNumbers[1] = 8; // valve 1 is on pin 8
valvePinNumbers[2] = 3; // valve 2 is on pin 3
```

 Whenever you need to look up some information based on an index, an array is a good way to do that. Think of it as a *lookup table*.

Now we can fix our question marks:

```
if ( ( now.hour() > onOffTimes[valve][onTime]) &&
     ( now.hour() < onOffTimes[valve][offTime]) ) {

    Serial.println("Turning valve ON");
    digitalWrite(valvePinNumbers[valve], HIGH);
}
else {
    Serial.println("Turning valve OFF");
    digitalWrite([valve], LOW);
}
```

One last thing: we're going to need to separate the four-digit numbers in the array into hours and minutes. It might be easier to do that when the user types in the information. We'll ask the user to add a : between the hours and minutes, we'll read them as separate numbers, do the conversion to minutes since midnight right there, and store the minutes since midnight in the array, as shown in Example 8-2.

You have probably already thought of two or three different ways of doing this. Most programming problems, in fact most engineering problems, can be solved many different ways. A professional programmer might consider efficiency, speed, memory usage, perhaps even cost, but as a beginner you should use whatever is easiest for you to understand.

Example 8-2. The expectValveSetting() function

```
/*
 * Read a string of the form "2N13:45" and separate it
 * into the valve number, the letter indicating ON or OFF,
 * and the time
 */
void expectValveSetting() {

  // The first integer should be the valve number
  int valveNumber = Serial.parseInt();

  // the next character should be either N or F
  char onOff = Serial.read();

  // next should come the hour
  int desiredHour = Serial.parseInt();

  // the next character should be ':'
  if (Serial.read() != ':') {
    Serial.println("no : found"); // Sanity check
    Serial.flush();
    return;
  }

  // next should come the minutes
  int desiredMinutes = Serial.parseInt();

  // finally expect a newline which is the end of
  // the sentence:
  if (Serial.read() != '\n') {
    Serial.println(
      "Make sure to end your request with a Newline"); // Sanity
```

```
check
    Serial.flush();
    return;
}


// Convert the desired hour and minute time
// to the number of minutes since midnight
int desiredMinutesSinceMidnight = (desiredHour*60 +
desiredMinutes);

// Now that we have all the information set it into the array
// in the correct row and column

if ( onOff == 'N') { // it's an ON time
    onOffTimes[valveNumber][ONTIME] = desiredMinutesSinceMidnight;
}
else if ( onOff == 'F') { // it's an OFF time
    onOffTimes[valveNumber][OFFTIME] =
desiredMinutesSinceMidnight;
}
else { // user didn't use N or F
    Serial.println(
        "You must use upper case N or F to indicate ON time or OFF
time");
    Serial.flush();
    return;
}

// now print the entire array so user can see what they set
for (int valve = 0; valve < NUMBEROFVALVES; valve++) {
    Serial.println();
    Serial.print("Valve ");
    Serial.print(valve);
    Serial.print(" will turn ON at ");
    Serial.print(onOffTimes[valve][ONTIME]);
    Serial.print(" and will turn OFF at ");
    Serial.print(onOffTimes[valve][OFFTIME]);
    Serial.println();
}
} // end of expectValveSetting()
```

Checking for Rain

What about checking for rain with the humidity sensor? You can do this at the same time you check the time, but it becomes a very long line. It's OK to use another `if()` statement; long-time programmers might tell you this is less efficient, but your garden won't care if the water comes on a split second later. Much more important is that you can read and understand the program.

 Be aware of well-intentioned experienced programmers who might show you clever tricks for reducing the number of lines in your program, or for improving efficiency. As you gain experience, it's good to understand why these tricks work, but as a beginner, you should always choose whatever is easiest for you to understand.

Example 8-3 shows a way to turn the water on only if it's not raining.

Example 8-3. Turn on the water only if it's not raining

```
if ( ( nowMinutesSinceMidnight >= onOffTimes[valve][ONTIME]) &&
     ( nowMinutesSinceMidnight < onOffTimes[valve][OFFTIME]) ) {
    // Before we turn a valve on make sure it's not raining
    if ( humidityNow > 50 ) { // Arbitrary; adjust as necessary
        // It's raining; turn the valve OFF
        Serial.print(" OFF ");
        digitalWrite(valvePinNumbers[valve], LOW);
    }
    else {
        // No rain and it's time to turn the valve ON
        Serial.print(" ON ");
        digitalWrite(valvePinNumbers[valve], HIGH);
    } // end of checking for rain
} // end of checking for time to turn valve ON
    else {
```

```
    Serial.print(" OFF ");
    digitalWrite(valvePinNumbers[valve], LOW);
}
```

Of course we'll make another function for this. Let's call it check-TimeControlValves().

We'll also create a separate function for reading the humidity sensor and the RTC. Let's call that getTimeTempHumidity().

Now our loop looks something like this:

```
void loop() {

    // Remind user briefly of possible commands
    Serial.println(
        "Type 'P' to print settings or 'S2N13:45' to set valve 2
ON time to 13:34");

    // Get (and print) the current date, time, temperature, and
humidity
    getTimeTempHumidity();

    // Check for new time settings:
    expectValveSettings();

    // Check to see whether it's time to turn any valve ON or OFF
    checkTimeControlValves();

    // No need to do this too frequently
    delay(5000);
}
```

Putting It All Together

We're almost done with the sketch. Just a couple of other minor things to consider, and then we can put this all together.

First, what if the user wants to see the current valve schedule? That's easy, but how does the user tell us? The user could type the letter *P* for "Print", but now the sketch needs to be ready for either the letter *P* or a number. This is tricky; it would be easier if we always expect the first character to be a letter and then decide whether to expect a number. Let's say that if the first letter is *P*, then we print the current settings, and if the first letter is *S*, then we expect a new setting to follow. If the user types anything other than *P* or *S*, we'd better remind them what's OK to type:

```
/*
 * Check for user interaction, which will be in the form of
something typed
 * on the serial monitor
 * If there is anything, make sure it's proper, and perform
the requested
 * action
 */
void checkUserInteraction() {
  // Check for user interaction
  while (Serial.available() > 0) {

    // The first character tells us what to expect for the
rest of the line
    char temp = Serial.read();

    // If the first character is 'P' then print the current
settings
    // and break out of the while() loop
    if ( temp == 'P') {
      for (int valve = 0; valve < NUMBEROFVALVES; valve++) {
        Serial.print("Valve ");
        Serial.print(valve);
        Serial.print(" scheduled ON at ");
        Serial.print(onOffTimes[valve][ONTIME]);
        Serial.print(" and OFF at ");
        Serial.print(onOffTimes[valve][OFFTIME]);
        Serial.println();
      }
      Serial.flush();
      break;
    } // end of printing current settings

    // If first character is 'S' then the rest will be a
setting
    else if ( temp == 'S') {
      expectValveSetting();
    }

    // Otherwise, it's an error. Remind the user what the
choices are
    // and break out of the while() loop
```

```
    else
    {
      printMenu();
      Serial.flush();
      break;
    }
  } // end of processing user interaction
}
```

The following code is the **printMenu()** function. It's short, but we might want to use it elsewhere. Also, in my experience, the menu tends to grow as the project becomes more and more complex, so this function is actually a nice way to document the menu within the sketch. For instance, later you might want to add a menu item to set the RTC time:

```
void printMenu() {
  Serial.println("Please enter P to print the current
settings");
  Serial.println("Please enter S2N13:45 to set valve 2 ON time
to 13:34");
}
```

 Any time a block of code is to be used more than once, it is a good candidate for becoming a function, no matter how short it is.

Finally, Example 8-4 shows the entire sketch.

Example 8-4. The irrigation system sketch

```
#include <Wire.h>    // Wire library, used by RTC library
#include "RTClib.h"  // RTC library
#include "DHT.h"     // DHT temperature/humidity sensor library

// Analog pin usage
const int RTC_5V_PIN = A3;
const int RTC_GND_PIN = A2;

// Digital pin usage
const int DHT_PIN  = 2;      // temperature/humidity sensor
const int WATER_VALVE_0_PIN = 8;
const int WATER_VALVE_1_PIN = 7;
```

```
const int WATER_VALVE_2_PIN = 4;

const int NUMBEROFVALVES = 3; // How many valves we have
const int NUMBEROFTIMES = 2;  // How many times we have

// Array to store ON and OFF times for each valve
// Store this time as the number of minutes since midnight
// to make calculations easier
int onOffTimes [NUMBEROFVALVES][NUMBEROFTIMES];
int valvePinNumbers[NUMBEROFVALVES];

// Which column is ON time and which is OFF time
const int ONTIME = 0;
const int OFFTIME = 1;

#define DHTTYPE DHT11
DHT dht(DHT_PIN, DHTTYPE); // Create a DHT object

RTC_DS1307 rtc;    // Create an RTC object

// Global variables set and used in different functions
DateTime dateTimeNow; // to store results from the RTC
float humidityNow;       // to store humidity result from the
DHT11 sensor

void setup(){

  // Power and ground to RTC
  pinMode(RTC_5V_PIN, OUTPUT);
  pinMode(RTC_GND_PIN, OUTPUT);
  digitalWrite(RTC_5V_PIN, HIGH);
  digitalWrite(RTC_GND_PIN, LOW);

  // Initialize the wire library
  #ifdef AVR
    Wire.begin();
  #else
    Wire1.begin(); // Shield I2C pins connect to alt I2C bus on
Arduino Due
  #endif

  rtc.begin();       // Initialize the RTC object
  dht.begin();       // Initialize the DHT object
  Serial.begin(9600); // Initialize the Serial object

  // Set the water valve pin numbers into the array
```

```
  valvePinNumbers[0] = WATER_VALVE_0_PIN;
  valvePinNumbers[1] = WATER_VALVE_1_PIN;
  valvePinNumbers[2] = WATER_VALVE_2_PIN;

};

void loop() {

  // Remind user briefly of possible commands
  Serial.println(
  "Type 'P' to print settings or 'S2N13:45' to set valve 2 ON
  time to 13:34");

  // Get (and print) the current date, time, temperature, and
humidity
  getTimeTempHumidity();

  // Check for request from the user
  checkUserInteraction();

  // Check to see whether it's time to turn any valve ON or OFF
  checkTimeControlValves();

  // No need to do this too frequently
  delay(5000);
}


/*
 * Get, and print, the current date, time, humidity, and
temperature
 */
void getTimeTempHumidity() {
  // Get and print the current time
  dateTimeNow = rtc.now();

  if (! rtc.isrunning()) {
    Serial.println("RTC is NOT running!");
    // use this to set the RTC to the date & time this sketch was
compiled
    // use this ONCE and then comment it out
    // rtc.adjust(DateTime(__DATE__, __TIME__));
    return; // if the RTC is not running don't continue
  }

  Serial.print(dateTimeNow.hour(), DEC);
```

```
Serial.print(':');
Serial.print(dateTimeNow.minute(), DEC);
Serial.print(':');
Serial.print(dateTimeNow.second(), DEC);

// Get and print the current temperature and humidity
humidityNow = dht.readHumidity();
// Read temperature as Celsius
float t = dht.readTemperature();
// Read temperature as Fahrenheit
float f = dht.readTemperature(true);

// Check if any reads failed and exit early (to try again).
if (isnan(humidityNow) || isnan(t) || isnan(f)) {
  Serial.println("Failed to read from DHT sensor!");
  return; // if the DHT is not running don't continue;
}

Serial.print(" Humidity ");
Serial.print(humidityNow);
Serial.print("% ");
Serial.print("Temp ");
Serial.print(t);
Serial.print("C ");
Serial.print(f);
Serial.print("F");
Serial.println();
} // end of getTimeTempHumidity()


/*
 * Check for user interaction, which will be in the form of
something
 * typed on the serial monitor
 * If there is anything, make sure it's proper, and perform the
requested
 * action
 */
void checkUserInteraction() {
  // Check for user interaction
  while (Serial.available() > 0) {

    // The first character tells us what to expect for the rest
of the line
    char temp = Serial.read();
```

```
    // If the first character is 'P' then print the current
settings
    // and break out of the while() loop
    if ( temp == 'P') {
      for (int valve = 0; valve < NUMBEROFVALVES; valve++) {
        Serial.print("Valve ");
        Serial.print(valve);
        Serial.print(" scheduled ON at ");
        Serial.print(onOffTimes[valve][ONTIME]);
        Serial.print(" and OFF at ");
        Serial.print(onOffTimes[valve][OFFTIME]);
        Serial.println();
      }
      Serial.flush();
      break;
    } // end of printing current settings

    // If first character is 'S' then the rest will be a setting
    else if ( temp == 'S') {
      expectValveSetting();
    }

    // Otherwise, it's an error. Remind the user what the choices
are
    // and break out of the while() loop
    else
    {
      printMenu();
      Serial.flush();
      break;
    }
  } // end of processing user interaction
}


/*
 * Read a string of the form "2N13:45" and separate it
 * into the valve number, the letter indicating ON or OFF,
 * and the time
 */
void expectValveSetting() {

  // The first integer should be the valve number
  int valveNumber = Serial.parseInt();

  // the next character should be either N or F
```

```
  char onOff = Serial.read();

  // next should come the hour
  int desiredHour = Serial.parseInt();

  // the next character should be ':'
  if (Serial.read() != ':') {
    Serial.println("no : found"); // Sanity check
    Serial.flush();
    return;
  }

  // next should come the minutes
  int desiredMinutes = Serial.parseInt();

  // finally expect a newline which is the end of
  // the sentence:
  if (Serial.read() != '\n') {
    Serial.println(
      "Make sure to end your request with a Newline"); // Sanity
check
    Serial.flush();
    return;
  }

  // Convert the desired hour and minute time
  // to the number of minutes since midnight
  int desiredMinutesSinceMidnight = (desiredHour*60 +
desiredMinutes);

  // Now that we have all the information set it into the array
  // in the correct row and column

  if ( onOff == 'N') { // it's an ON time
    onOffTimes[valveNumber][ONTIME] = desiredMinutesSinceMidnight;
  }
  else if ( onOff == 'F') { // it's an OFF time
    onOffTimes[valveNumber][OFFTIME] =
desiredMinutesSinceMidnight;
  }
  else { // user didn't use N or F
    Serial.println (
      "You must use upper case N or F to indicate ON time or OFF
time");
    Serial.flush();
    return;
```

```
    }

    // now print the entire array so user can see what they set
    for (int valve = 0; valve < NUMBEROFVALVES; valve++) {
      Serial.println();
      Serial.print("Valve ");
      Serial.print(valve);
      Serial.print(" will turn ON at ");
      Serial.print(onOffTimes[valve][ONTIME]);
      Serial.print(" and will turn OFF at ");
      Serial.print(onOffTimes[valve][OFFTIME]);
      Serial.println();
    }
} // end of expectValveSetting()


void checkTimeControlValves() {

    // First figure out how many minutes have passed since
midnight, since
    // we store ON and OFF time as the number of minutes since
midnight.
    // The biggest number
    // will be at 2359 which is 23 * 60 + 59 = 1159 which is less
    // than the maximum that can be stored in an integer so an "int"
    // is big enough
    int nowMinutesSinceMidnight =
      (dateTimeNow.hour() * 60) + dateTimeNow.minute();

    // Now check the array for each valve
    for (int valve = 0; valve < NUMBEROFVALVES; valve++) {
    Serial.print("Valve ");
      Serial.print(valve);
      Serial.print(" is now ");
      if ( ( nowMinutesSinceMidnight >= onOffTimes[valve][ONTIME])
&&
          ( nowMinutesSinceMidnight < onOffTimes[valve]
[OFFTIME]) ) {
        // Before we turn a valve on make sure it's not raining
        if ( humidityNow > 50 ) {
          // It's raining; turn the valve OFF
          Serial.print(" OFF ");
          digitalWrite(valvePinNumbers[valve], LOW);
        }
        else {
          // No rain and it's time to turn the valve ON
```

```
      Serial.print(" ON ");
      digitalWrite(valvePinNumbers[valve], HIGH);
    } // end of checking for rain
  } // end of checking for time to turn valve ON
  else {
    Serial.print(" OFF ");
    digitalWrite(valvePinNumbers[valve], LOW);
  }
  Serial.println();
  } // end of looping over each valve
  Serial.println();
}


void printMenu() {
  Serial.println("Please enter P to print the current settings");
  Serial.println("Please enter S2N13:45 to set valve 2 ON time to
13:34");
}
```

You can also download this sketch from from the example code link on the book's catalog page (*http://bit.ly/start_arduino_3e*).

Assembling the Circuit

Finally, we're done with the sketch and we've tested all the components! Are we ready to start soldering? Not quite: we've tested the various components separately, but not together. You might think that this step is unnecessary, but *integration testing* is vital. This step discovers unexpected interactions between components, whether hardware or software. For instance, two components might require the same feature on Arduino that is available on only one pin, or two libraries might conflict with each other, or the sketch logic might need to be reorganised. It's better to do this on the solderless breadboard in case wiring changes need to be made.

To get our full automatic garden-irrigation system, we need to combine three schematics: Figure 8-11, Figure 8-13, and the schematic for the RTC. Although our final system will have three valves, it will be quite a lot of work (and quite a squeeze for the solderless breadboard), for not much gain in information. If it works for one water valve, it should work for all three, so let's do just one valve for now.

Be very careful making assumptions, as they may be wrong and could come back to haunt you later. Never assume that if things work OK by themselves that they will work well together. Any engineer will tell you that integration testing is vital and very often finds problems that were not seen earlier.

Notice that I made an assumption: that testing with only one valve would be sufficient. This is exactly the sort of assumption I'm warning you against. For instance, more valves and more relays consume more power. Can the Arduino digital outputs provide power to all three relays if they are all on at the same time? Can the water valve power supply provide power to all three water valves if they are all on at the same time?

I've allowed myself to make these assumptions because I've done the rough calculations in my head, and because my years of experience tell me this is very low risk. However, as a beginner, you should avoid making such assumptions and test everything before you start soldering or assembling a project inside a case.

I've seen too many students have to take beautifully constructed projects apart because something didn't work the way they expected it to. (Even worse, it always seems to be the part that's hardest to get to.)

—Michael

Figure 8-15 shows the schematic and Figure 8-16 the pictorial circuit diagram for the entire system for one valve. Again, I've indicated the polarity of the water valve power supply and of the water valve, but this is relevant only if you have a DC system. If you have an AC system (which seems to be more common), these have no polarity.

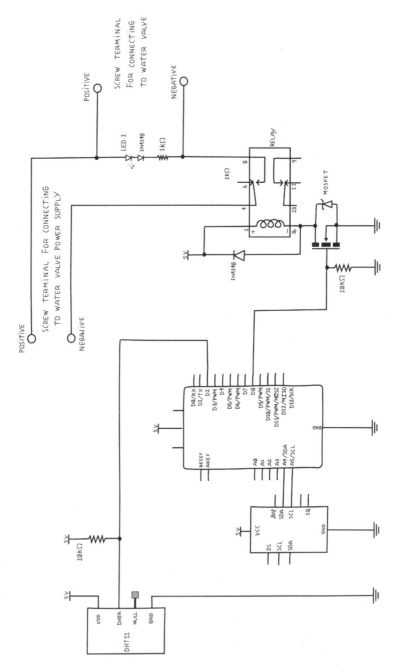

Figure 8-15. *Circuit schematic for full automatic garden-irrigation system, one valve*

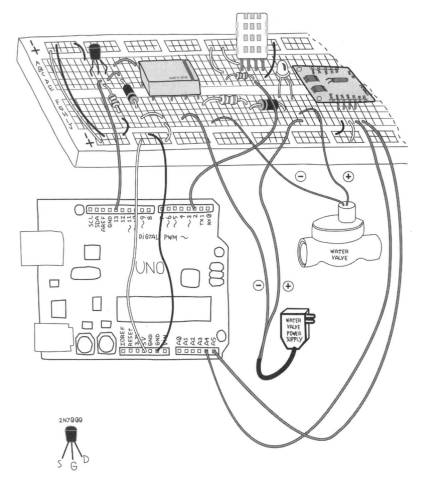

2N7000
S G D

Figure 8-16. *Pictorial circuit diagram for full automatic garden-irrigation system, one valve*

Before building a complex circuit, print out the schematic. As you build your circuit, use a coloured pen or highlighter to mark each connection you make. This will make it easy to see what you've done and what you've not yet done.

This is also useful when verifying a circuit, marking each connection as you verify it.

Build this on your breadboard and upload the complete system sketch from Example 8-4. It's OK that the sketch assumes three valves and we have only one: you can set times and allow the other two to activate, but nothing will happen.

Now test it: Press *P* to display the current settings and verify that they are all zero. Note the current time. Press *S* and set an ON time that is a minute or so away, and then set an OFF time for a minute after that. Your relay should click and the LED should come on. Your water valve may or may not do anything, depending on whether it works without any water pressure (mine makes a very reassuring click even without water).

Problems? Double-check your wiring. Pay particular attention to the diodes, the MOSFETs, and the relay. Remember that each pin of the MOSFET has a particular function, and you must use the correct pin. Remember that the diode is polarized. The black band indicates the cathode. Remember that the relay has a black strip indicating pins 8 and 9. If your water valve power supply and water valves use direct current (DC), make sure the positive and negative ends are connected where they are supposed to be. Check out Chapter 9.

This step also helps remind you of the importance of getting the MOSFETs, diodes, and relays wired up the right way. Once you solder these components, it won't be so easy to change. So once you get things working, make sure you understand why. Make notes of any mistakes you made and how you fixed them. You might even want to take some pictures of your breadboard for reference. It's always a good idea to document your work.

Once you're happy with that, you can move on to the Proto Shield.

The Proto Shield

As I mentioned earlier, we'll use the Proto Shield because it provides a secure and easy way to connect a project to Arduino. You can buy this from the Arduino Store (*http://arduino.cc/ protoShield*). There are many other Proto Shields available. Any will work, but you may have to make changes to the layout to fit your particular shield. Some shields come with the male headers, while others require that you buy them separately.

As you can see, the shield has male headers on the bottom that will plug into your Arduino, bringing all the Arduino pins to the shield. The shield has holes next to each Arduino pin for soldering wires, which are connected by the shield to the Arduino pin. To make a connection to an Arduino pin, simply solder a wire into the corresponding hole. This makes a much more reliable connection than poking the wires in to the headers as we've done in the past.

Most of the shield is taken up by a grid of tiny holes. They are a little like the holes on the solderless breadboard, in that you can place components and wires anywhere (almost) that you like, but unlike the solderless breadboard, very few connections are provided. You will be making most of the connections by soldering wires directly to the components, usually on the bottom of the board. You can minimize the number of connections you have to wire yourself by making clever use of any busses or other connected holes the shield offers.

When using a Proto Shield, or in fact any perforated soldering breadboard, it is common to put the components and wires on top and do the soldering on the bottom. This is especially important with a Proto Shield, as the bottom of the shield will be quite close to the Arduino, and there isn't much room there. Remember that none of your connections on the bottom of the shield must touch any metal on top of the Arduino, such as components, traces, or the USB port.

If you do have to place any components or wires on the bottom, make sure they are as flat as possible.

Laying Out Your Project on the Proto Shield

The first step is to think about what needs to fit on here, and where they will be. We need to make room for the MOSFETs, relays, LEDs, and screw terminals. The screw terminals should be along a side that is accessible, and it would be nice if the LEDs were near the appropriate screw terminals. The MOSFETs are pretty small and can go anywhere, but it would be nice if each were next to the relay it controlled.

The relays are the biggest objects, so we need to give them space before we fill up the shield.

It's easy to confuse the top of the shield and the bottom, so make sure you place the components on the right side. I have written TOP and BOTTOM on the following illustrations to remind you. I suggest you write TOP and BOTTOM on your shield with a permanent marker.

Make sure you don't use any holes that already have a function, such as the ICSP connector or the lone ground not far from the IOREF pin. In the illustrations, I have indicated these holes with a black-filled circle.

Avoid the area that sits above the USB connector. If you are using the Arduino Proto Shield, this area is intentionally free of holes.

Whenever you are soldering a circuit, think about where you will place things before you start any soldering. Start with connectors and the big items, and then place the smaller components close to where they need to connect. You can use the leads of the components to make the connections by bending the leads over on the underside of the shield and soldering them directly to the correct pins.

Don't solder anything until you are happy with the placement. Document your placement either by drawings or photographs before you begin soldering, in case anything falls out before you solder it into place.

I promised I'd explain what the sockets are for. You can see that the relays would have to be soldered into the Proto Shield. What happens if one of the relays goes bad? Happily, the relays will fit a socket. The socket gets soldered onto the Proto Shield, and the relay plugs into the socket.

If the relays get sockets, why doesn't every component get a socket? For a couple of reasons: resistors are easy to remove by desoldering. In the worst case, they can be cut out. Same thing for the MOSFETs. The relays would be very hard to desolder because they have eight pins. By the time we heat up the second pin, the first would have cooled down already. Also, once the relay is soldered in place, it's impossible to cut it out. Finally, the relay is a mechanical device with moving parts, and moving parts are more likely to fail than purely electronic parts. (Still, the relay should work for many years.)

Note that the sockets have an orientation: there is a small semicircle in the plastic indicating the top, or where pin 1 goes. The socket really doesn't care; it would work either way. This is meant to help you put the component in the right way around, so make sure you put the socket in facing the way you intend to wire it up. Again, making drawings and notes to yourself will help you later. Remember that when you flip the shield over, the orientation of the sockets will be reversed. I like to draw a circle around pin 1 of each socket on the bottom side of the board, to make sure I'm oriented properly.

When you flip the shield over, the sockets will fall out, so bend the leads over to hold them in place. They can be bent almost all the way as long as they don't touch any other holes.

 Whenever you are soldering a circuit, use sockets for relays and chips.

Figure 8-17 shows one possible layout.

Note that I have distorted the image a little to enlarge certain areas. We'll be doing lots of work there later, and I wanted to make it easier to see the details. The number of holes and the orientation of the rows and columns is accurate.

As we add the smaller components, I'll show you a trick. We'll make use of their leads to make some of our connections.

Look at the schematic. You'll see that the three diodes that go near the relay go from pin 1 to pin 16. If we place the diodes on that end, we can just bend the leads over on the underside of the shield and solder them directly to the correct pins. Make sure you observe the polarity of the diodes or you'll be cursing later. (I know because I've done that many times too.) The cathode is indicated by the ring near one end of the diode, and it goes to pin number 1 of the socket.

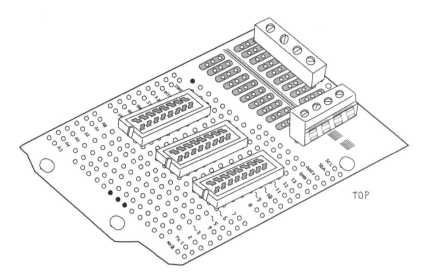

Figure 8-17. *One possible way to place the large components on the Proto Shield (note the orientation of the relay sockets)*

Bending the leads of the diode also helps keep it in place when you flip over the shield to solder the bottom.

The MOSFET has one pin (the drain) that is connected to the relay pin 16. Let's place that right next to the diode, and we can bend the MOSFET lead over and solder it to the diode. The 10 K ohm resistor that connects the gate of each MOSFET to the GND can go between the gate and the source of the MOSFET, since the source also goes to GND. We do this by standing the resistor on its end and using the resistor's leads to make the necessary connections, without having to add any wire.

Try to get all components as close to the shield as possible. The diodes should lie flat against the shield. You can bend the leads of the MOSFETs a little to make them low, but don't bend them too much or they will break. The resistor is standing, but one end of the resistor should be sitting on the shield.

I'll show you how to make all these connections in a moment.

Figure 8-18 shows the top view with the relay sockets, MOSFETs, diodes, and resistors added.

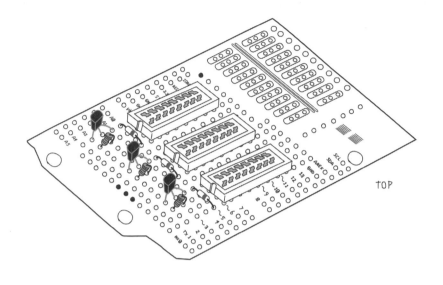

2N7000

S G D

Figure 8-18. *Top view with the relay sockets, MOSFETs, diodes, and resistors added*

What about the RTC and DHT11? The DHT11 needs to be out in the garden on four long wires. Rather than solder these wires directly to the Proto Shield, we'll solder a male header to the wire and mount a female header on the shield so that we can unplug it if necessary. I'll show you how to do this later. The 10 K ohm resistor (on the data pin of the DHT11) can be made to fit almost anywhere, so leave that for later.

The RTC already has male headers, so another female header will be perfect for the RTC. Remember that the RTC takes up quite a bit of space, so place this somewhere where there is room. The top edge of the board, after the MOSFETs and their related circuitry, might be a good place. I placed them in the very last row. This still left me an empty row between the MOS-

FETs and the female headers for any related wiring, as shown in Figure 8-19.

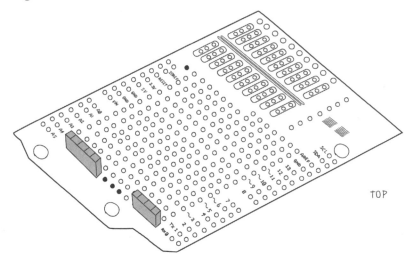

TOP

Figure 8-19. *A four position female headers for the DHT11 sensor and a five position female header for the RTC*

 Whenever you have to attach long wires from elsewhere to a board, don't solder the wires directly to the board. Instead, use a connector of some sort to make it easy to remove. A pair of male and female headers of the right number of positions make a good, inexpensive choice for small wires; screw terminals are good for larger wires.

Whenever you have to attach a module with headers to a board, don't solder the module headers directly to your board. Instead, mount a corresponding header of the other gender on your board. This will allow you to remove the module in case you need to for any reason.

Better add all these headers to the shopping list! These headers usually come in long strips and often multiple pieces. They are designed to be cut to whatever length you need. When you cut the male headers, you can usually break the strip exactly where you need it. When you cut the female headers, you have to sacrifice one position. Here are the additions that take our shopping list to version 0.5:

- Add set of female headers, .1" pitch, such as Adafruit product ID 598
- Add set of male headers, .1" pitch, such as Adafruit product ID 392

Soldering Your Project on the Proto Shield

For a great tutorial on how to solder, study the "Adafruit Guide to Excellent Soldering." (*http://bit.ly/solder-guide*)

Now, finally, you're ready to start soldering!

Don't rush. Be careful. Remember to breathe and relax. Double-check the schematic for each connection before you solder it. Inspect your work often for bad solder joints or other mistakes.

Don't try to make all the connections at once. Do them in small groups, and take a short break between groups. When you come back, double-check what you just did.

Try not to follow my directions blindly. Try to understand what's being done and make sure you agree with it.

You can remove the screw terminals and the female headers for now so that you can place the shield flat on your work surface to keep the parts in place.

Solder the sockets in place first to keep them from falling out. This will also give you a chance to practice soldering a little.

Next come the sets of diode, MOSFET, and resistor. Remember that we wanted to use their leads (on the bottom of the board) to make the connections. Bend them over very flat against the board so that the components are pulled tight against the board, and then over to where you need to make the connection. You don't need to wrap the lead around the pin; it's enough that the lead overlaps the pad around the pin. Make sure the solder flows onto all leads that are being connected.

When you're done soldering a connection, cut any excess leads as close as possible to the solder joint. You don't want any extra bit poking out that might touch something else later. This is detailed in the "Making a good solder joint" (*http://bit.ly/ 1tgWVHY*) tutorial by Adafruit.

When you trim a lead or wire after soldering, make sure you catch the piece you cut off and throw it away, to make sure it does not fall back on your work, where it might touch something it shouldn't and make a short circuit.

The MOSFET source pins all go to GND, so they chould be connected together. You can use the long leads of the resistors to connect all three together, which will provide a useful GND strip for anything else that needs to be connected to GND. These common strips are often called *buses*. Note that at this point our GND bus is not yet connected to the Arduino GND. I usually leave that for later so as not to occupy a hole that I later might need, but we have to remember to make this connection at the end.

Figure 8-20 shows the bottom view, with all the leads folded over and soldered in place. The shadowed areas indicate the relay sockets on the top side, and the cones indicate the solder.

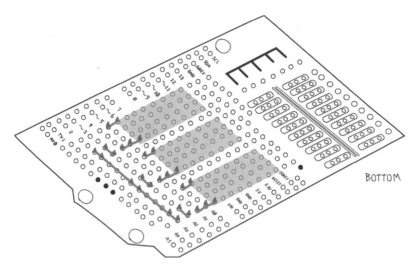

Figure 8-20. *Bottom view of the relay socket, MOSFET, and diode*

Now you can replace the screw terminals. Before you solder them in place, make sure the openings for the wire are facing the right direction, towards the outside of the board! (Another mistake I learned the hard way.) As always, make sure they are flush against the shield. Leave the female headers for later.

At this point it's a good idea to decide which screw terminal is what. Document this as shown in Figure 8-21 so that you don't forget and make the wrong connection later.

Figure 8-21. *Documenting which screw terminal does what*

Now you can add the indicator LEDs and their associated diodes and resistors. Again, by cleverly placing these components, you can use the component leads to make the connections. Note that this particular Proto Shield has some rows of three connected pins. I used these to help with the connections. Remember that LEDs and diodes are polarized: the anode (longer lead) of each LED goes to the screw terminal, and the anode of each diode goes the cathode of its LED.

It's important to note that there are two strips (buses) of 5V and GND that we're not using, so you have to carefully make sure the LED leads are above these and do not touch them. If the LED accidentally touched 5V or GND, this could bring the 24 V from the water valve power supply into Arduino, which would probably damage the Arduino. To be safe, you can cut a piece of electrical tape to the proper size and tape it over the 5V and GND buses. These buses are on the bottom side as well and must be avoided there too.

In Figure 8-22 we've shown the components quite high so that you can see where they are all connected, but when you build the circuit, place them as close to the shield as possible, as mentioned earlier. Remember that this illustration is distorted to enlarge certain details, but that the holes used are accurate.

In this and many of the following figures I've left off most of the components from previous steps so that you can see more clearly the components and locations of each step.

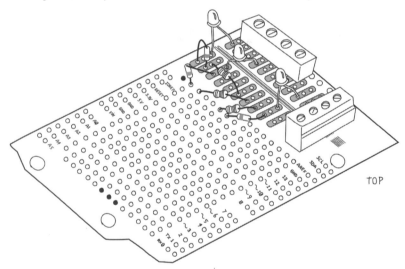

Figure 8-22. *Top view showing LED, resistor, and diode placement*

Figure 8-23 shows the bottom view. As before, we've used the component leads to make connections to each other and to the screw terminals. It's a good thing we documented the screw terminals, because we'd get confused now as to which one does what.

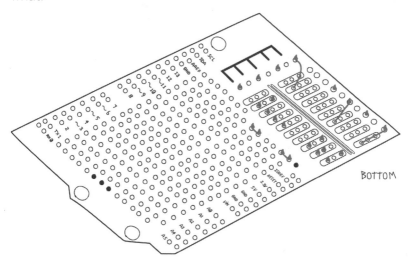

Figure 8-23. *Bottom view showing LED leads soldered to screw terminal pins*

Note the black circle near IOREF. That's a ground connection and must not be used unless you're making a connection to ground.

This is a complicated section. Study it carefully and don't solder anything until you are sure you understand what it's doing and are convinced it's going to the right place.

Now that all the components are on the shield, you have to wire up the remaining connections; use 22 AWG solid-core wire. Smaller would work as well but might be harder to work with. Use whatever works for you.

Choose a consistent color scheme: use red wire for anything connecting to 5V and black wire for anything connecting to ground. You can make up your own colour coding for the others, but don't use red or black for anything else. You might want to use orange for the water valve power supply positive connections, and green for the water valve power supply negative connections. Any wires that connect together should be the same colour, and any wires that do not connect together should be different colours.

The general principle is this: wires go on the top, and into holes that are next to the pin you need to connect to. On the bottom, you fold the wire over and solder to the appropriate pin, just as you did earlier with the component leads.

All the Arduino pins have their own holes, so you don't need to fold the wire over. Just solder the wire in the appropriate hole.

Sometimes you simply can't get close to the pin you need from the top. In these cases, it's OK to put the wires on the bottom, but make sure to keep them as flat as possible.

Let's start with the MOSFETs and their related circuitry. We've made all the connections we could with the leads. We need to connect the pin number 1 of all the relays to 5V. I've added red wires to make the 5V connection as shown in Figure 8-24 (remember the GND connection we'll make at the end).

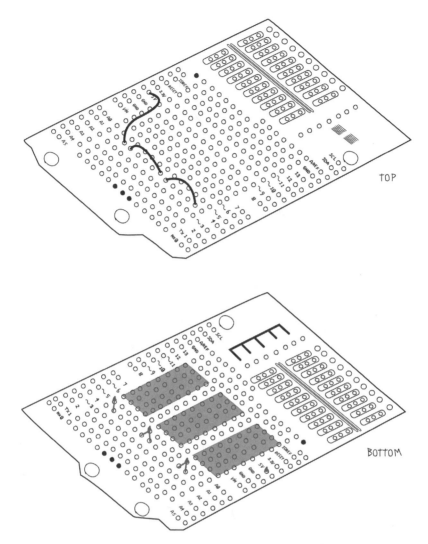

Figure 8-24. *Making the 5V connections to the relay circuitry*

Next, connect all the positive screw terminals. This is all done on the bottom of the board, as shown in Figure 8-25. Make sure not to allow any leads or solder to touch those 5V and GND buses!

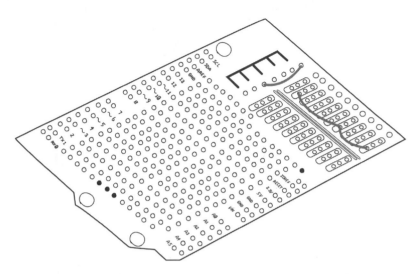

Figure 8-25. *Connecting all the positive screw terminals*

Connect the LED/resistor/diode chain to the negative screw terminal. Keep the middle hole clear, as we'll use it in the next step to connect to the relays. In this case, I used two wires on the top and one on the bottom, as shown in Figure 8-26.

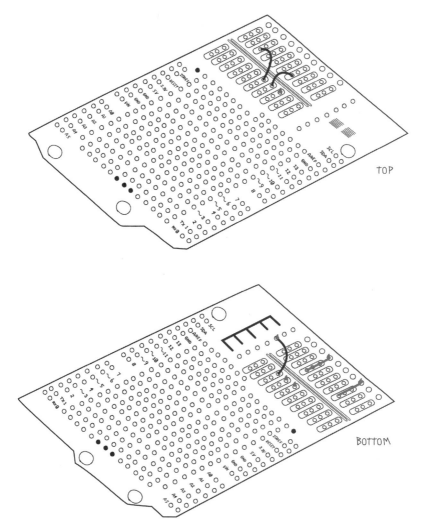

TOP

BOTTOM

Figure 8-26. *Connecting the negative screw terminals*

Now connect pin 8 of each relay to the appropriate negative screw terminal, as shown in Figure 8-27.

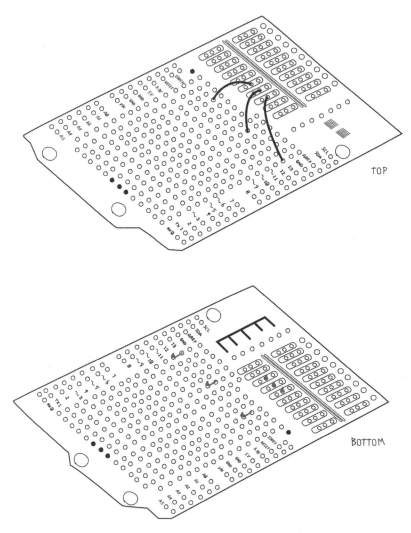

Figure 8-27. *Connecting pin 8 of each relay to the appropriate negative screw terminal*

Pin 4 of all the relays connect to the negative screw terminal of the water valve power supply. This is done with two wires on the top and one on the bottom, as shown in Figure 8-28.

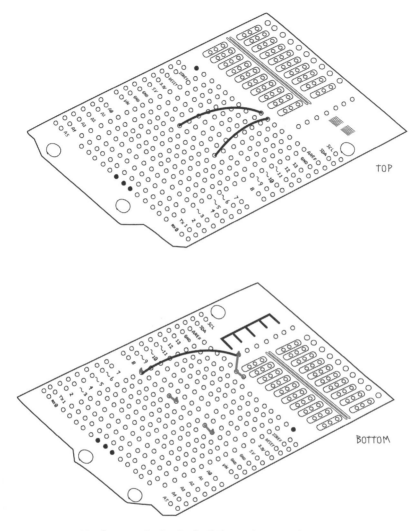

Figure 8-28. *Connect pin 4 of all the relays to the water valve power supply's negative screw terminal*

Next, connect the Arduino digital pins to the MOSFET gates, as shown in Figure 8-29. Remember the holes next to the Arduino pins are already connected to the Arduino pins so you don't need to bend the wire and solder it directly to the pin.

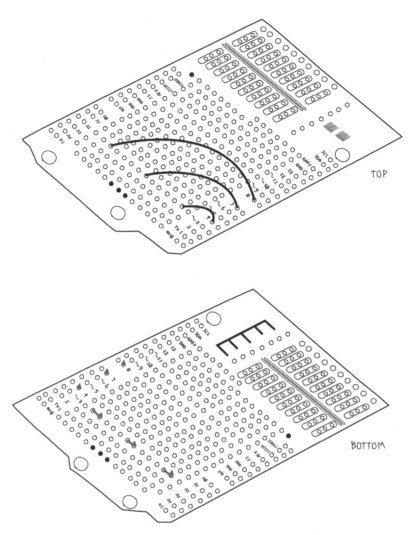

Figure 8-29. *Connecting the Arduino pins to the MOSFET gates*

Finally, add the two female headers: one for the RTC and one for the DHT11, and connect them to the appropriate Arduino pins. Don't forget the 10 K ohm resistor that the DHT11 needs, as shown in Figure 8-30. I've also taken this opportunity to connect the ground bus we created earlier to the Arduino GND pin.

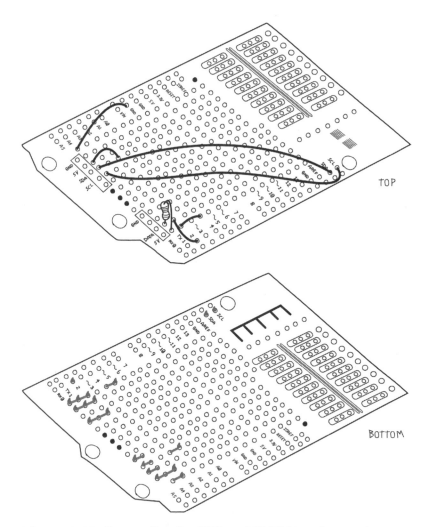

Figure 8-30. *Connecting the RTC and DHT11 headers to the Arduino pins*

Document which pins do what on the female headers, so that you can plug in the RTC and the DHT11 properly (an ultrafine Sharpie is handy for this).

The last step is to solder in the male or pass-through headers that will fit into the Arduino pins. This should be done last because the headers get in the way of the work you are doing on the bottom of the shield. Although you don't need all the pins, it's wise to put them all in for mechanical strength and possible future enhancements.

Don't forget that the male pins should point down; that is, towards the Arduino, as shown in Figure 8-34. Make sure the pins are straight so that they will fit into the female headers on your Arduino.

When you are done, it's time to test.

Testing Your Assembled Proto Shield

First test your shield without the valves or valve power supply. Plug the shield into your Arduino. Make sure the male header on the shield goes into the correct Arduino pins. Look between the two and make sure that no connections from the bottom of the shield are touching anything on the Arduino. If they are, you'll need to put some insulating electrical tape to prevent that from happening.

Connect your Arduino to a USB port on your computer and look for the Arduino ON LED. If it's off, it means that you have a short circuit, and that your computer has protected itself by turning off the USB port. Unplug the USB cable and find the problem before going further.

Next you can plug in and test the relays. Remember that they are polarized and that the stripe indicates pins 8 and 9. Upload the Blink example, each time testing a different relay. Test the relays one at a time to verify that each is working.

Next we'll test the water valve power supply and indicator LED. We'll do the valves last.

Connect the water valve power supply to the proper screw terminals. If the water valve system is a DC system, pay careful attention to the polarity.

Again upload the Blink example and test each relay in turn. This time, the appropriate LED should come on.

Now add the valves and check them, again using Blink for each valve.

Next let's check the RTC and the DHT11. Plug in the RTC to the header, making sure the RTC pins are in the right place. Use the RTC example to test.

Before you test the DHT11 sensor, add the long wires that will reach outdoors. Use the same colour wires as you did on the shield for consistency. Use stranded wires for this part because stranded wire is more flexible (Figure 8-31). For a more professional look, slip six short pieces of heat shrink tubing over the wires (two on each wire) before you solder. After you solder the wires to the headers and sensor, slide the tubing over the solder joints and shrink the tubing to hold it in place. I like to use clear tubing so that I can see if a solder joint breaks, and the clear tubing seems to be more in the spirit of "open".

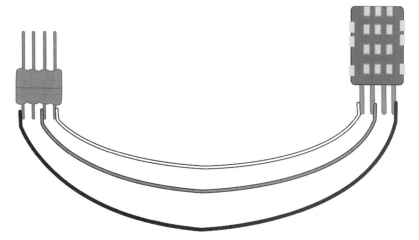

Figure 8-31. *Adding long stranded wires and a male header to the DHT11 sensor*

Solid-core wire should be used only in places where it will never move, i.e., when both ends are soldered to the same board.

Stranded wire should always be used between things that move relative to each other, such as between boards, or from a board to a connector.

I have seen many projects fail due to broken solid core wires that were moved too often.

Plug the male header from the DHT11 sensor into the corresponding female header, again paying careful attention to getting the right pins in the right place. Test using the DHT test example.

Assembling Your Project into a Case

Now we need to consider mounting the project in a case. Your best bet is a case that is not too deep to allow easy access, and lay everything out with a bit of room around them. Remember that the Arduino has a shield on top of it, and perhaps the RTC is standing vertically, adding to its height. The Arduino should be mounted on little feet called *standoffs*. A screw holds the Arduino to the standoff, and another screw from the back of the case will hold the standoff to the case, as shown in Figure 8-32.

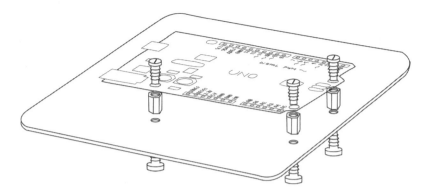

Figure 8-32. *Arduino mounted on standoffs inside case*

Always plan for a bigger case than you think you'll need. Don't forget the power supplies and connectors, and remember that wires take up space as well. You want to route the wires between the boards and not over the boards so they don't interfere if you have to work on or remove anything. For tidy wiring, I like these self-adhesive mounts for cable ties, as shown in Figure 8-33.

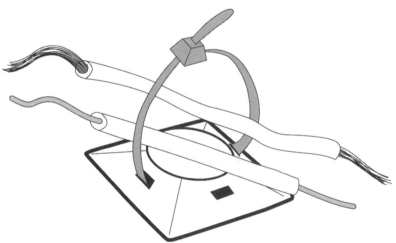

Figure 8-33. *Self-adhesive cable-tie mount with wires*

For projects with multiple power supplies, as we have here, consider mounting a small outlet strip inside the case. Use a strong double-sided tape to mount the outlet strip. If both your power supplies have only two prongs, you can use a two-pronged extension with at least two outlets instead of a larger three-pronged outlet strip.

This means another revision to our shopping list (we're now at revision 0.6):

- Case
- Standoffs
- Mounting screws and/or nuts
- Cable ties
- Adhesive cable-tie mounts (avilable from Jameco (*http://bit.ly/15p5rj5*))
- Strong double-sided tape (e.g., Digi-Key (*http://www.digi key.com*) part number M9828-ND)
- Outlet strip

I like to keep the power supplies away from the Arduino. You should mount the Arduino near the bottom so that the wires from the valves and the DHT11 sensor can enter from the bottom, but not too close. You'll be happy to have room for your hands and a screwdriver when attaching wires to the screw terminals, or if you need to work on the Arduino.

The USB cable can come out of any side.

You should always try to route the wires in straight lines and in tidy bundles. This makes working on the project later so much easier.

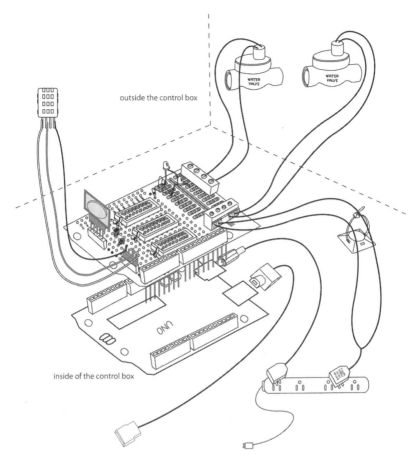

outside the control box

inside of the control box

Figure 8-34. *Completed automatic garden-irrigation system*

In this figure I've left many components off the Proto Shield to make it easier to see how things are connected. I've shown one cable tie, but you should use as many cable ties as you need to keep the wiring organized and tidy. Always use a cable tie before a wire leaves the enclosure; this serves as strain reliever in case the cable gets tugged: it will stress the cable-tie mount, and not your delicate circuitry.

As before, pay careful attention to the polarity of the water valve power supply, if it's a DC system.

Testing the Finished Automatic Garden Irrigation System

 Always test your projects in modules individually at first, in whatever way the project allows you to do this.

Start by testing the Arduino and Proto Shield without the two power supplies connected. This means that your computer is providing power to the Arduino. As before, use the Blink example to test each digital output. The LEDs won't light without the water valve power supply, but you should hear the relay clicking. Use DHTtester to test the DHT11 sensor, and the *ds1307* example to test the RTC.

This might seem like you're duplicating the tests you made after you assembled the Proto Shield, but there is a reason for this: before going on, you want to make sure that none of the work you've done has affected what worked before.

Now plug in the Arduino power supply and unplug your laptop to make sure your Arduino is getting powered without your laptop. Leave one of the relays clicking (using the Blink example) so that you can hear that your Arduino is still running the sketch.

Finally, connect the water valve power supply, upload the real program, and test the valves as you did earlier, by setting three different times in the near future. Check that each LED comes at the appropriate time, and that the water valves open and water flows.

Now relax in your garden and enjoy a well-deserved rest. You've accomplished a lot!

Congratulations! That was a complicated project! Take some pictures of your project, brag about it on Facebook, and submit it to the Arduino blog (*http://bit.ly/1rcG2Tz*).

Things to Try on Your Own

This is a complex project with many different components. There are endless things you can do differently. Here are a few suggestions:

• Modify the program to allow turning the water on and off at multiple times in one day.

• Add the day of the week, and allow for different schedules on different days of the week.

• Add an LED indicating that no ON and OFF times have been set. Turn this LED on upon reset, and turn it off the first time a time is set. This is useful in case your Arduino loses power and resets, in which case it will forget any ON and OFF settings you made.

• Add a small LCD display to report the current time and settings

• A more advanced exercise: The RTC module we used has a tiny memory chip in it as well that will not forget when it loses power. You can research how to use this, and save the valve ON and OFF settings into that memory, so that in case the Arduino resets, you will still have the settings.

Irrigation Project Shopping List

For your convenience, here is the final shopping list, with links:

• One Real Time Clock (RTC) (*http://bit.ly/11t1Huu*)

• One DHT11 Temperature and Humidity Sensor (*http://bit.ly/1HEHXFh*)

• One Arduino Proto Shield (*http://arduino.cc/protoShield*)

• Three electric water valves (*http://bit.ly/1AOXVee*)

• One transformer or power supply for the water valves (*http://bit.ly/1uCkwCt*)

• Three relays to control the water valves (*http://bit.ly/1xAspKl*)

• Three sockets for relays (*http://bit.ly/1zqJrNA*)

- Three LEDs as valve activation indicators (*http://ardu ino.cc/GreenLED*)

- Three 1 k resistors for LEDs (*http://arduino.cc/resis tor220Ohm*)

- One power supply for Arduino (*http://arduino.cc/powerSup ply9V*)

- Three MOSFETs to control the relays, 2N7000, 10-pack (*http://bit.ly/1uUkAm6*)

- Four resistors, 10 K ohm, 10-pack (*http://arduino.cc/resis tor10k*)

- Six diodes, 1N4148 or equivalent, 25-pack (*http://bit.ly/ 1xSkxbf*)

- Three relays (e.g., DS2E-S-DC5V, such as Digi-Key (*http:// www.digikey.com*) part number 255-1062-ND)

- Four dual-screw terminals (e.g., Jameco part no. 1299761 (*http://bit.ly/1ycB6P7*))

- Female headers, .1" pitch, 5-pack of 20 pin strips (*https:// www.adafruit.com/product/598*)

- Male headers, .1" pitch, 10-pack of 36 pin strips (*https:// www.adafruit.com/product/392*)

- Case: make your own or use a plastic storage bin (e.g., a Sterilite (*http://bit.ly/1vKO2Pl*) or a fancy metal box (available from Automation 4 Less (*http://bit.ly/1HEIg38*))

- Standoffs (*http://bit.ly/1yST9bq*)

- Mounting screw, standoff to case (*http://bit.ly/11t4zrj*)

- Mounting screw, Arduino to standoff (*http://bit.ly/1uUkTOf*)

- Cable ties

- Adhesive cable-tie mounts (*http://bit.ly/15p5rj5*)

- Strong double-sided tape (e.g., Digi-Key (*http://www.digi key.com*) part number M9828-ND)

- Outlet strip or extension cord with at least two outlets (*http://bit.ly/1pjR4Fm*)

9/Troubleshooting

There will come a moment in your experimentation when nothing will be working and you will have to figure out how to fix it. Troubleshooting and debugging are ancient arts in which there are a few simple rules, but most of the results are obtained through careful work and paying attention to details.

The most important thing to remember is that you have not failed! Most makers, both amateurs and professionals, spend most of their time fixing mistakes that they themselves have made. (True, we get better at finding and fixing problems, but we also create more complicated problems.)

As you work more with electronics and Arduino, you too will learn and gain experience, which will ultimately make the process less painful. Don't be discouraged by the problems that you will find—it's all easier than it seems at the beginning. The more mistakes you make and correct, the better you will get at finding them.

As every Arduino-based project is made both of hardware and software, there will be more than one place to look if something goes wrong. While looking for a bug, you should operate along three lines: understanding, simplification and segmentation, and exclusion and certainty.

Understanding

Try to understand as much as possible how the parts that you're using work and how they're supposed to contribute to the finished project. This approach will allow you to devise some way to test each component separately. If you've not already done so, try drawing a schematic of your project. This helps you understand your project, and is also useful if you have to ask for help. Schematics are discussed in Appendix D.

Simplification and Segmentation

The ancient Romans used to say *divide et impera*: divide and rule. Try to break down (mentally) the project into its components by using the understanding you have and figure out where the responsibility of each component begins and ends.

Exclusion and Certainty

While investigating, test each component separately so that you can be absolutely certain that each one works by itself. You will gradually build up confidence about which parts of a project are doing their job and which ones are dubious.

Debugging is the term used to describe this process as applied to software. The legend says it was used for the first time by Grace Hopper back in the 1940s, when computers were mostly electromechanical, and one of them stopped working because actual insects got caught in the mechanisms.

Many of today's bugs are not physical anymore: they're virtual and invisible, at least in part. Therefore, they require a sometimes lengthy and boring process to be identified. You will have to trick the invisible bug into revealing itself.

Debugging is a little like detective work. You have a situation that you need to explain. To do this, you do some experiments and come up with results, and from these results you try to deduce what has caused your situation. It's elementary, really.

Testing the Arduino Board

Before trying very complicated experiments, it's wise to check the simple things, especially if they don't take much time. The first thing to check is that your Arduino board works, and the very first example, Blink, is always a good place to start, because you are probably most familiar with it, and because the LED that is already on your Arduino means that you won't depend on any external components.

Follow these steps before you connect your project to your Arduino. If you've already connected jumpers between your

Arduino and your project, remove them for now, keeping careful track of where each jumper should go.

Open the basic Blink example in the Arduino IDE and upload it to the board. The onboard LED should blink in a regular pattern.

What if Blink doesn't work?

Before you start blaming your Arduino, you should make sure that a few things are in order, as airline pilots do when they go through a checklist to make sure that the airplane will be flying properly before takeoff:

- Is your Arduino getting power, either from a computer though a USB cable or from an external power supply? If the green light marked PWR turns on, this means that your Arduino is getting power. If the LED seems very faint, something is wrong with the power.

  If you are using a computer, make sure the computer is on (yes, it sounds silly, but it has happened). Try a different USB cable. Inspect the computer's USB port and the Arduino's USB plug to see whether there is any damage. Try a different USB port on your computer, or a different computer entirely. If you have lots of USB cables on your workbench, make sure that the one plugged into your Arduino is the one that is plugged into the computer (yes, we've done this).

  If you are using external power, verify that the external power supply is plugged in. Make sure your outlet strip or extension cord is plugged in. If you are using an outlet strip with a switch, make sure it's turned on.

  (If you are using a very old Arduino, verify that the power selection jumper is in the correct position. Modern Arduinos do this automatically and don't have this jumper.)

- If the Arduino is brand new, the yellow LED marked *L* might start blinking even before you upload the Blink example. This is likely the test program that was loaded at the factory to test the board and is OK. It's also OK if there is no blinking—it simply means a different test program is running on it.

- Verify that the sketch uploaded successfully.

  If upload failed, check first that your program has no errors by clicking Verify.

  Make sure you selected the proper board in the Tools menu. As you start to accumulate different Arduino boards, it's a good habit to always make sure that the board selected is indeed the one you have connected.

  Check that the port in the Tools menu is selected properly. If you unplugged your Arduino at some point, it might appear on a different port.

  Sometimes you have to unplug the Arduino and plug it in again. If you have the Serial Port selection menu open, you have to close it (just move to another tab) and then go back to Tools→Serial Port and select the proper port.

  Try uploading again. On rare occasions, an upload will fail for no apparent reason, and will succeed the next time without changing anything.

  Poor-quality USB cables can sometimes prevent the driver from finding the Arduino Uno. If your Arduino port doesn't show up in the Port list, try using a known good USB cable.

Once you have the basic Blink example loaded and the LED blinking, you can be confident that your Arduino has basic functionality, and can proceed to the next step.

Testing Your Breadboarded Circuit

The next step is to check for short circuits between 5V and GND on your project. Connect your Arduino to your breadboard by running a jumper from the 5V and GND connections to the positive and negative rails of the breadboard. (Notice we are following the "divide and rule" principle by connecting only these two jumpers, and not all the jumpers for your project.) If the green PWR LED turns off, remove the jumpers immediately. This means there is a big mistake in your circuit and you have a "short circuit" somewhere. When this happens, your board draws too much current and the power gets cut off to protect the computer.

If you're concerned that you may damage your computer, remember that almost all computers limit the amount of current a USB device can draw. If the device tries to take too much current, the computer immediately disables power on the USB port. Also, the Arduino board is fitted with a *polyfuse*, a current-protection device that resets itself when the fault is removed.

If you're really paranoid, you can always connect the Arduino board through a self-powered USB hub. In this case, if it all goes horribly wrong, the USB hub is the one that will be pushing up daisies, not your computer.

If you're getting a short circuit, you have to start the "simplification and segmentation" process. What you must do is go through every sensor and actuator in the project and connect just one at a time until you identify the part or connection that is causing the short circuit.

The first thing to start from is always the power supply (the connections from 5V and GND). Look around and make sure that each part of the circuit is powered properly. The most likely cause is a jumper that is in the wrong place. Other causes might be an incorrect component such as a resistor with too small a value, or a switch or transistor that is connecting 5V to GND. Less likely but also possible is a piece of wire or a screw that happens to be touching both 5V and GND somewhere.

> Working step by step and making one single modification at a time is the number one rule for fixing stuff. This rule was hammered into my young head by my school professor and first employer, Maurizio Pirola. Every time I'm debugging something and things don't look good (and believe me, it happens a lot), his face pops in my head saying, "One modification at a time...one modification at a time" and that's usually when I fix everything. This is very important, because

you will know what fixed the problem. (It's all too easy to lose track of which modification actually solved the problem, which is why it's so important to make one at a time.)

—Massimo

Each debugging experience will build up in your head a "knowledge base" of defects and possible fixes. And before you know it, you'll become an expert. This will make you look very cool, because as soon as a newbie says, "This doesn't work!" you'll give it a quick look and have the answer in a split second.

Isolating Problems

Another important rule is to find a reliable way to reproduce a problem. If your circuit behaves in a funny way at random times, try really hard to identify what seems to cause this. Does it happen only when you press a switch? Only when an LED lights up? Whenever you move a jumper? (Many problems are caused by loose wires, either not connecting where they should, or connecting where they shouldn't.) Try to repeat the steps that cause the problem, paying attention to small details and making one change at a time: does it happen every time the LED lights up, or only if you press the switch while the LED is on? This process will allow you to think about a possible cause. It is also very useful when you need to explain to somebody else what's going on.

Describing the problem as precisely as possible is also a good way to find a solution. Try to find somebody to explain the problem to—in many cases, a solution will pop into your head as you articulate the problem. Brian W. Kernighan and Rob Pike, in *The Practice of Programming* (Addison-Wesley, 1999), tell the story of one university that "kept a teddy bear near the help desk. Students with mysterious bugs were required to explain them to the bear before they could speak to a human counselor." If you don't have someone (or a teddy bear) to talk to, start writing an email describing your problem.

Problems Installing Drivers on Windows

Sometimes the Found New Hardware Wizard fails to locate the proper driver. In this case you might have to manually tell it where the driver is located.

The Found New Hardware Wizard will first ask you whether to check Windows Update; select the "No, not at this time" option and click Next.

On the next screen, choose "Install from a list or specific location" and click Next.

Navigate to and select the Uno's driver file, named *ArduinoUNO.inf*, located in the *Drivers* folder of the Arduino Software download (not the *FTDI USB Drivers* subdirectory). Windows will finish up the driver installation from there.

Problems with the IDE on Windows

If you get an error when you double-click the Arduino icon, or if nothing happens, try double-clicking the *Arduino.exe* file as an alternative method to launch Arduino.

Windows users may also run into a problem if the operating system assigns a COM port number of COM10 or greater to Arduino. If this happens, you can usually convince Windows to assign a lower port number to Arduino by freeing up (temporarily) a COM port with a lower number.

First, open up the Device Manager by clicking the Start menu, right-clicking Computer (Vista) or My Computer (XP), and choosing Properties. On Windows XP, click Hardware and choose Device Manager. On Vista, click Device Manager (it appears in the list of tasks on the left of the window).

Look for the serial devices in the list under Ports (COM & LPT). Find a serial device that you're not using that is numbered COM9 or lower. A modem or serial port make good candidates. Right-click it and choose Properties from the menu. Then, choose the Port Settings tab and click Advanced. Set the COM port number to COM10 or higher, click OK, and click OK again to dismiss the Properties dialogue.

Now, do the same with the USB Serial Port device that represents Arduino, with one change: assign it the COM port number (COM9 or lower) that you just freed up.

Identifying the Arduino COM Port on Windows

Connect your Arduino Uno to your computer via a USB cable.

Open the Device Manager by clicking the Start menu, right-clicking Computer (Vista) or My Computer (XP), and choosing Properties. On Windows XP, click Hardware and choose Device Manager. On Vista, click Device Manager (it appears in the list of tasks on the left of the window).

Look for the Arduino device in the list under Ports (COM & LPT). The Arduino will appear as Arduino UNO and will have a name like COM7, as shown in Figure 9-1.

 On some Windows machines, the COM port has a number greater than 9; this numbering creates some problems when Arduino is trying to communicate with it.

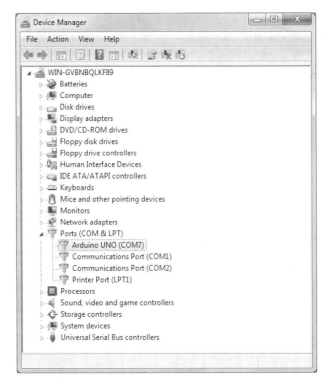

Figure 9-1. *The Windows Device Manager showing all available serial ports*

Other Debugging Techniques

- Ask someone else to look over your project. We sometimes become blind to our own mistakes. Don't tell the other person what connections you meant to make; have them verify that you have correctly implemented whatever diagram you are working from. This way, you don't prejudice them to see what you intended to do and miss the mistake. (If you don't have a diagram, you should make one.)

- "Divide and rule" works for sketches, too. Save a copy of your sketch, and then start removing the parts of your sketch that don't have anything to do with the part that is giving you trouble. You might find an unexpected interraction between something that seems to be working fine and the problem. If this doesn't solve your problem, it will provide the minimal test program that demonstrates your problem, which will make it easier when you ask for help.

- If your project involves any sensors (including switches), test each one individually with the most basic appropriate examples: AnalogReadSerial and DigitalReadSerial, which you can find at *File Examples 01.Basics AnalogReadSerial/ DigitalReadSerial.*

- If any sensor fails, verify that the Arduino input is working correctly. Disconnect your sensor, and connect a jumper from the suspect input directly to 5V and GND (one at a time, obviously), while monitoring with AnalogReadSerial or DigitalReadSerial. You should see 0 when the input is connected to GND, and 1 or 1023 when the input is connected to 5V.

  If you have multiple sensors and one is failing, swap parts of the circuit (one at a time) between the one that works and the one that fails, and see if the problem moves.

- If your project involves any actuators, test each one individually with the most basic appropriate examples: Blink or Fade. If the actuator fails, replace it with an LED to make sure the Arduino output is working correctly.

- If your sketch involves decision making, such as `if` statements, use the `Serial.println()` function to tell you what it's doing. This is also useful in loops, to make sure the loop is stopping when you think it should.

- If you are using any libraries, verify they work correctly using the examples that came with them. If you are having trouble with a library that is not from Arduino, see if there is a forum or other online community for that library and join it.

If these suggestions don't help, or if you're having a problem not described here, check out the Arduino troubleshooting page (*http://bit.ly/1zVIIWG*).

How to Get Help Online

If you are stuck, don't spend days running around alone—ask for help. One of the best things about Arduino is its community. You can always find help if you can describe your problem well.

Get into the habit of cutting and pasting things into a search engine and see whether somebody is talking about it. For example, when the Arduino IDE spits out a nasty error message, copy and paste it into a Google search and see what comes out. You may have to put the message in quotes to prevent matching those words in random orders. Do the same with bits of code you're working on or just a specific function name. If you get too many hits that aren't useful, add the word *Arduino* to the search.

Look around you: everything has been invented already and it's stored somewhere on a web page. I'm surprised how often something I think happened only to me turns out to be well documented on the Web, along with the solution.

For further investigation, start from the main website (*http:// www.arduino.cc*) and look at the FAQ (*http:// www.arduino.cc/en/Main/FAQ*); then move on to the Arduino Playground (*http://playground.arduino.cc/*), a freely editable wiki that any user can modify to contribute documentation. It's one of the best parts of the whole open source philosophy: people contribute documentation and examples of anything you can do with Arduino. Before you start a project, search the Playground and you'll find a bit of code or a circuit diagram to get you started.

(Consider paying the open source community back, by documenting a project you came up with or a solution you found that was not previously documented.)

If you still can't find an answer that way, search the Arduino forum (*http://forum.arduino.cc/*).

After you've tried everything else, it's time to post a question to the Arduino forum. Pick the correct board for your problem: there are different areas for software or hardware issues and even forums in different languages. If you're unsure which board is appropriate, post in the Project Guidance board.

Compose your post carefully. Post as much information as you can, and be clear and thorough. Taking the time to clearly and correctly describe your problem is well worth it. This also shows that you've already done as much as you could by yourself, and aren't relying on the forum to do your work for you. Here are some guidelines:

- Read the post titled "How to use this forum—please read" (*http://bit.ly/1y9jh2u*).

- What Arduino board are you using?

- What operating system are you using to run the Arduino IDE?

- What version of the Arduino IDE are you using?

- Give a general description of what you're trying to do. Post links to data sheets of strange parts you're using. Don't clutter up your post with irrelevant information, such as the project concept or a picture of the enclosure if it doesn't pertain to the problem.

- Post the minimal sketch and/or circuit (schematic diagrams are great for this) that shows your problem. (You found this when you were debugging, right?). The "How to use this forum" post shows you how to format code and include attachments.

- When you search the forum for existing help, pay attention to the culture, especially the types of questions that get good help versus the types of questions that don't. You want to copy the style of those that work.

- Describe exactly what you think should happen, and what is happening instead. Don't just say, "It doesn't work." If you get an error message, post the error. If your program prints output, post that output.

- Now that you've described your problem carefully, go back and revise the subject. You want a subject that summarises the technical issue, not the goal of your project (e.g., "pressing multiple switches causes short circuit" and not "help with control panel for rocket ship").

Remember that the number of answers you get, and how quickly you get them, depends on how well you formulate your question.

Your chances increase if you avoid these things at all cost (these rules are good for any online forum, not just Arduino's):

- Typing your message all in CAPITALS. It annoys people a lot and is like walking around with "newbie" tattooed on your forehead (in online communities, typing in all capitals is considered "shouting").
- Posting the same message in several different parts of the forum.
- *Bumping* your message by posting follow-up comments asking, "Hey, how come no one replied?" or even worse, simply posting the text "bump." If you didn't get a reply, take a look at your posting. Was the subject clear? Did you provide a well-worded description of the problem you are having? Were you nice? Always be nice.
- Writing messages like "I want to build a space shuttle using Arduino how do I do that." This means that you want people to do your work for you, and this approach is simply not fun for a real tinkerer. It's better to explain what you want to build and then ask a specific question about one part of the project and take it from there. In addition to helpful answers, you might also get useful suggestions for your larger project.
- A variation of the previous point is when the question is clearly something the poster of the message is getting paid to do. If you ask specific questions, people are happy to help, but if you ask them to do all your work (and you don't share the money), the response is likely to be less nice.
- Posting messages that look suspiciously like school assignments and asking the forum to do your homework. Professors like me roam the forums and slap such students with a large trout. Forum regulars are also good at spotting these.

A/The Breadboard

The process of getting a circuit to work might involve making lots of changes until it behaves properly. As you iterate your circuit, you might get ideas that help you refine your design, perhaps improving its behaviour, making it more reliable, or requiring fewer parts. The design evolves in your hands as you try different combinations; that's something like an electronic equivalent to sketching.

Ideally, you'd like a way to build circuits that allows you to change the connections between components quickly and easily. While soldering is great for creating reliable, permanent circuits, you'd like something faster.

The answer to this problem is a very practical device called a *solderless breadboard*. As you can see from Figure A-1, it's a small plastic board full of holes, each of which contains a spring-loaded contact. You can push a wire or a component's leg into one of the holes, and the spring will hold the component or wire in place. More important, because the spring is connected to adjacent springs, it will establish an electrical connection with certain other holes.

In the central region (the rows labeled a–j), the springs run vertically, and so any component placed in these holes is immediately connected with any other components in the same vertical column of holes.

Some solderless breadboards have additional rows: two on top and two on the bottom, often indicated by red and blue stripes and sometimes marked with + and −. These rows are connected horizontally, and are intended for any electrical signal that gets used often. These rows are perfect for 5V or GND, which are the most common connections in the projects in this book, and in almost any electronic project. These rows are often called *rails* or *buses*.

If you connect the red row (or the one marked +) to the 5V on your Arduino, and the blue (or the one marked −) row to the GND on your Arduino, you will always have 5V and GND near any point of the breadboard.

A good example of these rails is in Chapter 7.

 On some breadboards, the rails do not go all the way across, and instead are broken in the middle. Sometimes this is indicated by a break in the red or blue stripe, and sometimes this is indicated by a gap between pins that is slightly larger than usual. As it is easy to forget this, many people permanently leave a jumper bridging this break on each row.

Some components, like resistors, capacitors, and LEDs, have long flexible legs that can be bent to reach holes in different places.

However, other components, like chips, have legs (known to techies as *pins*) that cannot be moved. These pins almost always have a spacing of 2.54 mm, so the holes on the solderless breadboard use this same spacing.

Most chips have two rows of pins, and if the breadboard columns were connected all the way across, the pins on one side of the chip would be connected (by the breadboard) to the pins on the other side. This is the reason for the gap in the middle, which interrupts each vertical line of holes. If you place a chip so that it straddles the gap, the pins on one side will not be connected to the pins on the other side. Clever, eh?

 Some breadboards have letters indicating the rows, and numbers indicating the columns. We won't be referring to these, as not all breadboards are the same. Whenever we say pin number, we're referring to the Arduino pin, and not anything on the breadboard.

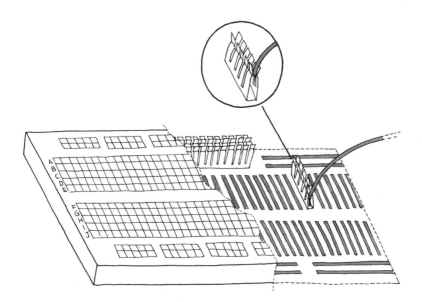

Figure A-1. *The solderless breadboard*

B/Reading Resistors and Capacitors

In order to use electronic parts, you need to be able to identify them, which can be a difficult task for a beginner. Most of the resistors that you find in a shop have a cylindrical body with two legs sticking out and have strange coloured markings all around them. When the first commercial resistors were made, there was no way to print numbers small enough to fit on their body, so clever engineers decided that they could just represent the values with stripes of coloured paint.

Today's beginners have to figure out a way to interpret these signs. The key is quite simple: generally, there are four stripes, and each colour represents a number. One of rings is usually gold-coloured; this one represents the tolerance of that resistor. To read the stripes in order, hold the resistor so the gold (or silver in some cases) stripe is to the right. Then, read the colours and map them to the corresponding numbers. In the following table, you'll find a translation between the colours and their numeric values.

Colour	Value
Black	0
Brown	1
Red	2
Orange	3
Yellow	4
Green	5
Blue	6
Violet	7
Grey	8
White	9

Colour	Value
Silver	10 %
Gold	5 %

For example, brown, black, orange, and gold markings mean 1 0 3 ffl5 %. Easy, right? Not quite, because there is a twist: the third ring actually represents the number of zeros in the value. Therefore 1 0 3 is actually 1 0 followed by three zeros, so the end result is 10,000 ohms ffl5 %.

Electronics geeks tend to shorten values by expressing them in kilo ohms (for thousands of ohms) and mega ohms (for millions of ohms), so a 10,000-ohm resistor is usually shortened to 10 K ohm, while 10,000,000 becomes 10 M ohm. Because engineers are fond of optimising everything, on some schematic diagrams you might find values expressed as 4k7, which means 4.7 kilo ohms, or 4700.

Sometimes you'll run into resistors with a higher precision of 1 or 2 %. These resistors add a fifth ring so that the value can be specified more precisely. It's the same code, but with the first three rings representing the value and the fourth ring representing the number of zeros after the value. The fifth ring would be the tolerance: red for 2 % and brown for 1 %. For example, the 10 K ohm example (brown, black, orange, and gold) would be brown, black, black, red, and brown, for a 1 % resistor.

Capacitors are a bit easier: the barrel-shaped capacitors (electrolytic capacitors) generally have their values printed on them. A capacitor's value is measured in farads (F), but most capacitors that you encounter will be measured in micro farads (μF). So if you see a capacitor labeled 100 μF, it's a 100 micro farad capacitor.

Many of the disc-shaped capacitors (ceramic capacitors) do not have their units listed, and use a three-digit numeric code indicating the number of pico farads (pF). There are 1,000,000 pF in one μF. Similar to the resistor codes, you use the third number to determine the number of zeros to put after the first two, with one difference: if you see 0–5, that indicates the number of zeros. 6 and 7 are not used, and 8 and 9 are handled differently.

If you see 8, multiply the number that the first two digits form by 0.01, and if you see 9, multiply it by 0.1.

So, a capacitor labeled 104 would be 100,000 pF or 0.1 µF. A capacitor labeled 229 would be 2.2 pF.

As a reminder, here are the multipliers commonly used in electronics.

Multiplier	Value	Example
M (mega)	10^6 = 1,000,000	1,200,000 ohm = 1.2 M ohm
k (kilo)	10^3 = 1,000	470,000 ohm = 470 K ohm
m (milli)	10^{-3} = .001	.01 A = 10 mA
u (micro)	10^{-6} = .000001	4700 u amps = 4.7 mA
n (nano)	10^{-9}	10,000 n farads = 10 µF
p (piclo)	10^{-12}	1,000,000 p f = 1 µF

C/Arduino Quick Reference

Here is a quick explanation of all the standard instructions supported by the Arduino language.

For a more detailed reference, see the Arduino "Language Reference" page (*http://bit.ly/1ycDNQA*).

Structure

An Arduino sketch runs in two parts:

```
void setup()
```

This is where you set things up that have to be done once before the loop starts running, and then don't need to happen again.

```
void loop()
```

This contains the main code of your sketch. It contains a set of instructions that get repeated over and over until the board is switched off.

Special Symbols

Arduino includes a number of symbols to delineate lines of code, comments, and blocks of code.

; (semicolon)

Every instruction (line of code) is terminated by a semicolon. This syntax lets you format the code freely. You could even put two instructions on the same line, as long as you separate them with a semicolon. (However, this would make the code harder to read.)

Example:

```
delay(100);
```

{} (curly braces)

These are used to mark blocks of code. For example, when you write code for the `loop()` function, you have to use a curly brace before and after the code.

Example:

```
void loop() {
    Serial.println("ciao");
}
```

Comments

These are portions of text ignored by the Arduino microcontroller, but are extremely useful to explain to others (and to remind yourself) what a piece of code does.

There are two styles of comments in Arduino:

```
// single-line: this text is ignored until the end of the
line
/* multiple-line:
    you can write
    a whole poem in here
*/
```

Constants

Arduino includes a set of predefined keywords with special values.

`HIGH` and `LOW` are used, for example, when you want to turn on or off an Arduino pin. `INPUT` and `OUTPUT` are used to set a specific pin to be either an input or an output.

`true` and `false` are used to test whether a condition or expression is true or false. They are used primarily with *comparison operators*.

Variables

Variables are named areas of the Arduino's memory where you can store data. Your sketch can use and manipulate this data by referring to it by the variable name. As the word *variable* suggests, variables can be changed as many times as you like.

Because Arduino is a very simple microcontroller, when you declare a variable, you have to specify its type. This means telling the microcontroller the size of the value you want to store.

Following are the *datatypes* that are available.

boolean
Can have one of two values: `true` or `false`.

char
Holds a single character, such as the letter *A*. Like any computer, Arduino stores it as a number, even though you see text. When chars are used to store numbers, they can hold values from –128 to 127. A char occupies 1 byte of memory.

 There are two major sets of characters available on computer systems: ASCII and UNICODE. ASCII is a set of 127 characters that was used for, among other things, transmitting text between serial terminals and time-shared computer systems such as mainframes and minicomputers. UNICODE is a much larger set of values used by modern computer operating systems to represent characters in a wide range of languages. ASCII is still useful for exchanging short bits of information in languages such as Italian or English that use Latin characters, Arabic numerals, and common typewriter symbols for punctuation and the like.

---

byte
Holds a number between 0 and 255. Like a char, a byte uses only 1 byte of memory. Unlike chars, a byte can store only positive numbers.

int
Uses 2 bytes of memory to represent a number between –32,768 and 32,767. The `int` is the most common datatype used in Arduino. If you are unsure of what datatype to use, try an int.

unsigned int

> Like int, uses 2 bytes of memory, but the unsigned prefix means that it can't store negative numbers, so its range goes from 0 to 65,535.

long

> This is twice the size of an int and holds numbers from −2,147,483,648 to 2,147,483,647.

unsigned long

> Unsigned version of long; it goes from 0 to 4,294,967,295.

float

> This is quite big and can hold floating-point values, which is a fancy way of saying that you can use a float to store numbers with a decimal point. A float will eat up 4 bytes of your precious RAM, and the functions that can handle them use up a lot of code memory as well, so use floats only when you need to.

double

> Double-precision floating-point number, with a maximum value of $1.7976931348623157 \times 10^{308}$. Wow, that's huge!

string

> A set of ASCII characters used to store textual information (you might use a string to send a message via a serial port, or to display on an LCD display). For storage, they use 1 byte for each character in the string, plus a null character (1 byte) at the end to tell Arduino that it's the end of the string. The following are equivalent:

```
char string1[]  = "Arduino"; // 7 chars + 1 null char
char string2[8] = "Arduino"; // Same as above
```

array

> A list of variables that can be accessed via an index. They are used to build tables of values that can easily be accessed. For example, if you want to store different levels of brightness to be used when fading an LED, you could create six variables called light01, light02, and so on. Better yet, you could use a simple array like this:

```
int light[6] = {0, 20, 50, 75, 100};
```

The word *array* is not actually used in the variable declaration: the symbols [] and {} do the job.

Arrays are ideal when you want to do the same thing to a whole lot of pieces of data, because you can write what you need to do once and then perform it on each variable in the array simply by changing the index—for example, using a `for` loop.

Variable Scope

Variables in Arduino have a property called *scope*. Variables can be local or global, depending on where they are declared.

A global variable is one that can be seen (and used) by every function in a program. Local variables are visible only to the function in which they are declared.

When programs start to get larger and more complex, local variables are a useful way to ensure that each function has access to its own variables. This prevents programming errors when one function inadvertently modifies variables used by another function. Variables that must be used by multiple functions can be global.

In the Arduino environment, any variable declared outside of a function (e.g., `setup()`, `loop()`, or your own functions), is a global variable. Any variable declared within a function is local (and accessible) only within that function.

It is also sometimes handy to declare and initialize a variable inside a `for` loop. This creates a variable that can only be accessed from inside the `for` loop braces. In fact, any time a variable is declared within *curly braces*, it is local only within that block of code.

Control Structures

Arduino includes keywords for controlling the logical flow of your sketch.

if...else
> This structure makes decisions in your program. `if` must be followed by a question specified as an expression contained

in parentheses. If the expression is true, whatever follows will be executed. If it's false, the block of code following else will be executed. The else clause is optional.

Example:

```
if (val == 1) {
   digitalWrite(LED,HIGH);
}
```

for

Lets you repeat a block of code a specified number of times.

Example:

```
for (int i = 0; i < 10; i++) {
     Serial.print("ciao");
}
```

switch case

The if statement is like a fork in the road for your program. *switch case* is like a massive roundabout. It lets your program take a variety of directions depending on the value of a variable. It's quite useful to keep your code tidy as it replaces long lists of if statements.

It's important to remember the **break** statement at the end of each **case**, or else Arduino will execute the instructions of the following cases, until it reaches a **break** or the end of the *switch case*.

Example:

```
switch (sensorValue) {
    case 23:
      digitalWrite(13,HIGH);
      break;
    case 46:
      digitalWrite(12,HIGH);
      break;
    default: // if nothing matches this is executed
      digitalWrite(12,LOW);
      digitalWrite(13,LOW);
}
```

while

Similar to if, this executes a block of code if a certain condition is true. However, if executes the block only once, whereas while keeps on executing the block as long as the condition is true.

Example:

```
// blink LED while sensor is below 512
sensorValue = analogRead(1);
while (sensorValue < 512) {
    digitalWrite(13,HIGH);
    delay(100);
    digitalWrite(13,HIGH);
    delay(100);
    sensorValue = analogRead(1);
}
```

do...while

Just like while, except that the code is run before the condition is evaluated. This structure is used when you want the code inside your block to run at least once before you check the condition.

Example:

```
do  {
    digitalWrite(13,HIGH);
    delay(100);
    digitalWrite(13,HIGH);
    delay(100);
    sensorValue = analogRead(1);
} while (sensorValue < 512);
```

break

This term lets you break out of a while or for loop even if the loop condition says to go on looping. It's also used to separate the different sections of a *switch case* statement.

Example:

```
// blink LED while sensor is below 512
do  {
    // Leaves the loop if a button is pressed
    if (digitalRead(7) == HIGH)
        break;
    digitalWrite(13,HIGH);
```

```
        delay(100);
        digitalWrite(13,LOW);
        delay(100);
        sensorValue = analogRead(1);
    } while (sensorValue < 512);
```

continue

> When used inside a loop, continue lets you skip the rest of the code inside it and force the condition to be tested again.

> Example:

```
for (light = 0; light < 255; light++)
{
  // skip intensities between 140 and 200
  if ((x > 140) && (x < 200))
    continue;
  analogWrite(PWMpin, light);
  delay(10);
}
```

> continue is similar to break, but break leaves the loop, while continue goes on with the next repetition of the loop.

return

> Stops running a function and returns to whatever called the function. You can also use this to return a value from inside a function.

> For example, if you have a function called computeTemperature() and you want to return the result to the part of your code that invoked the function, you would write something like this:

```
int computeTemperature() {
    int temperature = 0;
    temperature = (analogRead(0) + 45) / 100;
    return temperature;
}
```

Arithmetic and Formulas

You can use Arduino to make complex calculations using a special syntax. + and − work just like you've learned in school; multiplication is represented with an *, and division with a /.

There is an additional operator called *modulo* (%), which returns the remainder of an integer division.

Just as you learned in algebra, you can use as many levels of parentheses as you wish to to group expressions the proper way. Contrary to what you might have learned in school, square brackets and curly braces are not used for arithmetic forumulas because they are reserved for other purposes (array indexes and blocks, respectively).

Example:

```
a =  2 + 2;
light = ((12 * sensorValue) - 5 ) / 2;
remainder = 7 % 2; // returns 1
```

Comparison Operators

When you specify conditions or tests for if, while, and for statements, these are the operators you can use:

==	Equal to
!=	Not equal to
<	Less than
>	Greater than
<=	Less than or equal to
>=	Greater than or equal to

When testing for equality, be very careful to use the == comparison operator and not the = assignment operator, or your program will not behave the way you expect.

Boolean Operators

These are used when you want to combine multiple conditions. For example, if you want to check whether the value coming from a sensor is between 5 and 10, you would write this:

```
if ( (sensor => 5) && (sensor <=10) )
```

There are three Boolean operators: *and*, represented with &&; *or*, represented with | |; and finally *not*, represented with !.

Compound Operators

These are special operators used to make code more concise for some very common operations like incrementing a value.

For example, to increment `value` by 1, you would write:

```
value = value +1;
```

but using a compound operator, this becomes:

```
value++;
```

It's perfectly fine not to use these compound operators, but they are so common that, as a beginner, you will have a hard time learning from examples if you don't understand these operators.

increment and decrement (-- and ++)

These operators increment or decrement a value by 1. Be careful—they work both in front of or behind a variable, but they have a very subtle difference: if you write `i++`, this first increments `i` by 1 and then evaluates to the equivalent of `i + 1`, while `++i` first evaluates to the value of `i` and then increments `i`. The same applies to `--`.

`+=` , `-=`, `*=`, and `/=`
> Similar to `++` and `--`, but these allow you to increment and decrement by values other than 1, and also allow multiplication and division. The following two expressions are equivalent:
>
> ```
> a = a + 5;
> a += 5;
> ```

Input and Output Functions

One of the main jobs of Arduino is to input information from sensors and to output values to actuators. You've already seen some of these in the example programs throughout the book.

pinMode(pin, mode)
> (Re)configure a digital pin to behave either as an input or an output.

Example:

```
pinMode(7,INPUT); // turns pin 7 into an input
```

Forgetting to set pins to outputs using `pinMode()` is a common cause of faulty or nonfunctioning output.

Although typically used in `setup()`, `pinMode()` can be used in a loop as well if you need to change the pin's behaviour.

(When a function name is used in text, it is often written with empty parentheses at the end to indicate that a function is being discussed.)

digitalWrite(pin, value)
Turns a digital pin either HIGH or LOW. Pins must be explicitly made into an output using `pinMode()` before `digitalWrite()` will have the expected effect.

Example:

```
digitalWrite(8,HIGH); // sets digital pin 8 to 5 V
```

Note that while HIGH or LOW usually correspond to on and off, respectively, this depends on how the pin is used. For example, an LED connected between 5V and a pin will turn on when that pin is LOW and turn off when the pin is HIGH.

int digitalRead(pin)
Reads the state of an input pin, and returns HIGH if the pin senses some voltage or LOW if there is no voltage applied.

Example:

```
val = digitalRead(7); // reads pin 7 into val
```

int analogRead(pin)
Reads the voltage applied to an analogue input pin and returns a number between 0 and 1023 that represents the voltages between 0 and 5 V.

Example:

```
val = analogRead(0); // reads analog input 0 into val
```

analogWrite(pin, value)
Changes the PWM rate on one of the PWM pins. `pin` can only be a pin that supports PWM, that is, pins 3, 5, 6, 9, 10, or 11

on the Uno, and pins 3, 5, 6, 9, 10, 11, or 13 on the Leonardo. **value** must be a number between 0 and 255. You can think of **value** to represent the average amount of power Arduino will deliver, where a **value** of zero corresponds to fully off, while a **value** of 255 corresponds to fully on.

Example:

```
analogWrite(9,128); // Dim an LED on pin 9 to 50%
```

A **value** of 0 sets the output fully LOW, while a **value** of 255 sets an output fully HIGH.

shiftOut(dataPin, clockPin, bitOrder, value)
Sends data to a *shift register*, devices that are used to expand the number of digital outputs. This protocol uses one pin for data and one for clock. **bitOrder** indicates the ordering of bytes (least significant or most significant) and **value** is the actual data to be sent out.

Example:

```
shiftOut(dataPin, clockPin, LSBFIRST, 255);
```

unsigned long pulseIn(pin, value)
Measures the duration of a pulse coming in on one of the digital inputs. This is useful, for example, to read some infrared sensors or accelerometers that output their value as pulses of changing duration.

Example:

```
time = pulsein(7,HIGH); // measures the time the next
                        // pulse stays high
```

Time Functions

Arduino includes functions for measuring elapsed time and also for pausing the sketch.

unsigned long millis()
Returns the number of milliseconds that have passed since the sketch started.

Example:

```
duration = millis()-lastTime; // computes time elapsed
since "lastTime"
```

delay(ms)
Pauses the program for the amount of milliseconds specified.

Example:

```
delay(500); // stops the program for half a second
```

delayMicroseconds(µs)
Pauses the program for the given amount of microseconds.

Example:

```
delayMicroseconds(1000); // waits for 1 millisecond
```

Math Functions

Arduino includes many common mathematical and trigonometric functions:

min(x, y)
Returns the smaller of x and y.

Example:

```
val = min(10,20); // val is now 10
```

max(x, y)
Returns the larger of x and y.

Example:

```
val = max(10,20); // val is now 20
```

abs(x)
Returns the absolute value of x, which turns negative numbers into positive. If x is 5, it will return 5, but if x is −5, it will still return 5.

Example:

```
val = abs(-5); // val is now 5
```

constrain(x, a, b)

Returns the value of x, constrained between a and b. If x is less than a, it will just return a, and if x is greater than b, it will just return b.

Example:

```
val = constrain(analogRead(0), 0, 255); // reject values
bigger than 255
```

map(value, fromLow, fromHigh, toLow, toHigh)

Maps a value in the range `fromLow` and `maxLow` to the range `toLow` and `toHigh`. Very useful to process values from analogue sensors.

Example:

```
val = map(analogRead(0),0,1023,100, 200); // maps the
value of
                                          // analog 0 to a
value
                                          // between 100
and 200
```

double pow(base, exponent)

Returns the result of raising a number (**base**) to a value (**exponent**).

Example:

```
double x = pow(y, 32); // sets x to y raised to the 32nd
power
```

double sqrt(x)

Returns the square root of a number.

Example:

```
double a = sqrt(1138); // approximately 33.73425674438
```

double sin(rad)

Returns the sine of an angle specified in radians.

Example:

```
double sine = sin(2); // approximately 0.90929737091
```

double cos(rad)

Returns the cosine of an angle specified in radians.

Example:

```
double cosine = cos(2); // approximately -0.41614685058
```

double tan(rad)
Returns the tangent of an angle specified in radians.

Example:

```
double tangent = tan(2); // approximately -2.18503975868
```

Random Number Functions

If you need to generate random numbers, you can use Arduino's pseudorandom number generator. Random numbers are useful if you want your project to behave differently each time it's used.

randomSeed(seed)
Resets Arduino's pseudorandom number generator. Although the distribution of the numbers returned by `random()` is essentially random, the sequence is predictable. So, you should reset the generator to some random value. A good seed is a value read from an unconnected analogue input, as an unconnected pin will pick up random noise from the surrounding environment (radio waves, cosmic rays, electromagnetic interference from cell phones and fluorescent lights, etc.) and so will be unpredictable.

Example:

```
randomSeed(analogRead(5)); // randomize using noise from
pin 5
```

long random(max) long random(min, max)
Returns a pseudorandom `long` integer value between `min` and `max` - 1. If `min` is not specified, the lower bound is 0.

Example:

```
long randnum = random(0, 100); // a number between 0 and 99
long randnum = random(11);     // a number between 0 and 10
```

Serial Communication

As you saw in Chapter 5, you can communicate with devices over the USB port using a serial communication protocol. Following are the serial functions.

Serial.begin(speed)

Prepares Arduino to begin sending and receiving serial data. You'll generally use 9600 baud (bits per second) with the Arduino IDE serial monitor, but other speeds are available, usually no more than 115,200 bps. The specific baud rate doesn't matter much, as long as both sides agree and use the same rate.

Example:

```
Serial.begin(9600);
```

Serial.print(data) Serial.print(data, encoding)

Sends some data to the serial port. The encoding is optional; if not supplied, the data is treated as much like plain text as possible.

Examples (note that the final example uses `Serial.write`):

```
Serial.print(75);       // Prints "75"
Serial.print(75, DEC);  // The same as above.
Serial.print(75, HEX);  // "4B" (75 in hexadecimal)
Serial.print(75, OCT);  // "113" (75 in octal)
Serial.print(75, BIN);  // "1001011" (75 in binary)
Serial.write(75);       // "K" (the letter K happens
                        // to be 75 in the ASCII set)
```

Serial.println(data) Serial.println(data, encoding)

Same as `Serial.print()`, except that it adds a carriage return and linefeed (\r\n) as if you had typed the data and then pressed Return or Enter.

Examples:

```
Serial.println(75);       // Prints "75\r\n"
Serial.println(75, DEC);  // The same as above.
Serial.println(75, HEX);  // "4B\r\n"
Serial.println(75, OCT);  // "113\r\n"
Serial.println(75, BIN);  // "1001011\r\n"
```

int Serial.available()

Returns how many unread bytes are available on the serial port for reading via the **read()** function. After you have **read()** everything available, **Serial.available()** returns 0 until new data arrives on the serial port.

Example:

```
int count = Serial.available();
```

int Serial.read()

Fetches 1 byte of incoming serial data.

Example:

```
int data = Serial.read();
```

Serial.flush()

Because data may arrive at the serial port faster than your program can process it, Arduino keeps all the incoming data in a buffer. If you need to clear the buffer and let it fill up with fresh data, use the **flush()** function.

Example:

```
Serial.flush();
```

D/Reading Schematic Diagrams

In most of this book I've given very detailed illustrations to describe how to assemble the circuits, but as you can imagine, it's not exactly a quick task to draw one of those for each experiment you might want to document.

Similar issues arise, sooner or later, in every discipline. In music, after you write a nice song, you need to write it down using musical notation.

Engineers, being practical people, have developed a quick way to capture the essence of a circuit in order to be able to document it and later rebuild it or pass it to somebody else.

In electronics, *schematic diagrams* (or *schematics*) allow you to describe your circuit in a way that is understood by the rest of the community. Individual components are represented by *schematic symbols* that are a sort of abstraction of either the shape of the components or the essence of them. For example, the capacitor is made of two metal plates separated by either air or plastic; therefore, its symbol is as shown in Figure D-1.

Figure D-1. *Schematic symbol for a capacitor*

Another clear example is the inductor, which is built by winding copper wire around a cylindrical shape; consequently, the symbol looks like Figure D-2.

Figure D-2. *Schematic symbol for an inductor*

The connections between components are usually made using either wires or tracks on the printed circuit board and are represented on the diagram as simple lines. When two wires are connected, the connection is represented by a big dot placed where the two lines cross, as shown in Figure D-3.

Figure D-3. *Schematic symbol showing connected wires*

This is all you need to understand basic schematics. Figure D-4 shows schematic symbols for components that are commonly found in Arduino circuits.

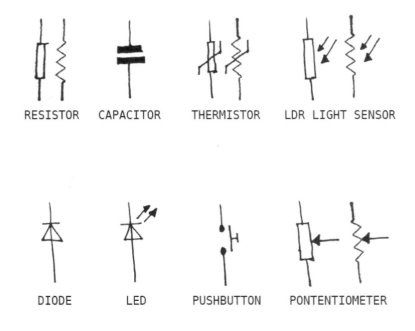

RESISTOR   CAPACITOR   THERMISTOR   LDR LIGHT SENSOR

DIODE   LED   PUSHBUTTON   PONTENTIOMETER

Figure D-4. *Common schematic symbols seen in Arduino circuits*

You may encounter variations in these symbols (for example, both variants of resistor symbols are shown here). See Wikipedia (*http://bit.ly/1zVpAa3*) for a larger list of electronics symbols.

In addition to this (somewhat) standard set of symbols, there are conventions for how schematics are organised. Schematics are drawn so that information flows from left to right. For example, a radio would be drawn starting with the antenna on the left, following the path of the radio signal as it makes its way to the speaker, which would be the last thing on the right.

Figure D-5 describes the pushbutton circuit shown earlier in this book.

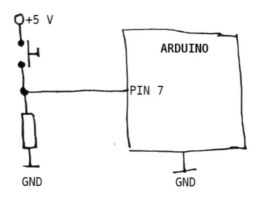

Figure D-5. *A pushbutton connected to an Arduino digital input*

You can see that the Arduino has been reduced to a box with a pin and GND, because these are the only important things to know about Arduino for this particular circuit. You can also see two wires that are shown connected the label GND. This means the wires are connected together. Connecting wires via labels is useful for connections that get very busy (such as GND) or have to get from one side of the schematic to the far side, crossing many other wires and components.

Chapter 8 shows many practical examples of schematics, and "Electronic Schematic Diagrams" on page 119 discusses schematic diagrams in a little more detail.

Index

Symbols

// (comment indicator), 32, 135
1N4007 diode, 74
22 AWG solid core wire, 175
2N7000 MOSFET, 117
5V pin, 42
; (semicolon), 46
= (assignment) operator, 44
== (comparison) operator, 44
{} (curly braces), 31

A

AC vs. DC electricity, 130
actuators, 25
    testing, 202
Adafruit, 105
    male headers at, 111
Adafruit Guide to Excellent Soldering, 170-184
Adafruit Guide To Making A Good Solder Joint, 171
airplane cockpit idea, 87
Alighieri, Dante, 32
Ampère, André-Marie, 39
Analog Devices TMP36, 70
analog inputs, 66-70
    controlling blinking LEDs with, 67
    controlling brightness of LEDs with, 68
    dealing with I2C communications, 114
    on Leonardo, 79
    pins, 17
analog pins, 17
    on Uno vs. Leonardo, 89
analogRead() function, 66, 70, 74
    return values of, 69

analogWrite() function, 57, 61
    Leonardo and, 78
anodes, 27, 59
Arduino
    groups, 106
    LEDs on, 30
    meetups, 106
    philosophy of, 5
    schematic symbol for, 120
    troubleshooting, 194
Arduino Esplora, 78
Arduino Forum, 203
    etiquette in, 205
Arduino Leonardo, 77-89
    Arduino Uno vs., 78
    connecting to Macs, 79
    I2C pins on, 114
    keyboard message example, 80-83
    LED controller on, 57
    microcontroller on, 77
    mouse control example, 83-87
    Mouse library, 87
    older models vs., 77
    pins, 89
    ports on, 87
    USB Keyboard library, 87
Arduino Library, 116
Arduino Micro, 78
Arduino platform, 15-23
    hardware, 15-17
Arduino robot, 78
Arduino Store, 41, 65, 102, 162
Arduino Uno, 16
    LED controller on, 57
    Leonard vs., 78
Arduino Yun, 78
AREF pins, 88, 114
arithmetic in Processing, 222

H

hackerspaces, 106
hacks
    keyboard, 11
    toy, 13
Haque, Usman, 13
hardware
    Proto Shield, 162-184
    to blink LEDs, 57
heat shrink tubing, 125
heat-dependent resistors, 69
Hedberg, Sara Reese, 12
help, finding, 203
HTML hexadecimal color codes, 94
humans, detecting, 56
hydraulic system, 38

I

I promessi sposi (Manzoni), 32
I2C port, 88
    RTCs and, 111
ICSP port, 88
IDII Ivrea, 2
if statements, 44, 219
    debugging, 202
Igoe, Tom, 74
IKEA FADO table lamp, 103
increment/decrement operators,
    224
information flows on electronics
    schematics, 120
input/output, 53-74
    analog inputs, 66-70
    complex sensors, 74
    digital I/O pins, 17
    functions for, 224
    homemade, 56
    light sensors, 64
    on/off sensors, 53
    powering devices and, 72
    pushbuttons, 40-44, 46-51
    serial, 70
    thermostat, 54
    toggle switches, 53
insulating tape, 125

int keyword, 33
Integrated Development Environ-
    ment (IDE), 15, 18
    installing, 19-23
    Macintosh, installing on, 19-21
    on Mac, 20
    on Windows, installing, 21
    opening, 28
    programming in, 31-36
    Serial Monitor in, 71
    specifying board for, 21, 79
    troubleshooting on Windows,
        199
integration testing, 158
Intelligent Systems and Their
    Applications, IEEE, 12
interaction design, 3
interactive devices, 25-51
    actuators, 25
    LEDs, controlling, 26-36
    sensors, 25
interactive lamp example, 36-51
    pushbuttons, 40-44
Internet forums, 14
    Arduino Forum, 203
    etiquette in, 205
interruptions, 7
    external, 89
IRF520 MOSFET, 73
Ivrea, 12

J

jumper wires, connecting TinyRTC
    with, 113
junk, 12

K

Kernighan, Brian W., 198
Keyboard library, 78
keyboard message example, 80-83
Keyboard Setup Assistant (Mac),
    79
Keyboard.print() function, 82
Keyboard.println() function, 82
keyboards, hacking, 80-83
Kurt, Tod E., 94

L

M

N

About the Authors

Massimo Banzi is the cofounder of the Arduino project and has worked for clients such as Prada, Artemide, Persol, Whirlpool, V&A Museum, and Adidas.

Michael Shiloh is an associate professor at the California College of the Arts, where he teaches electronics, programming, robotics, and mechatronics. Trained formally as an electrical engineer, Michael worked for various consumer and embedded engineering firms before discovering a passion for teaching. Michael also prefers applying his engineering skills to creative and artistic devices rather than consumer devices. Michael frequently lectures and speaks at conferences and universities worldwide. In 2013, Michael started working for Arduino, speaking about and teaching the open source electronics prototyping platform to new audiences.

Colophon

The cover image is by Judy Aime' Castro. The cover and text font is Myriad Pro; the heading font is Benton Sans; and the code font is Dalton Maag's Ubuntu Mono.